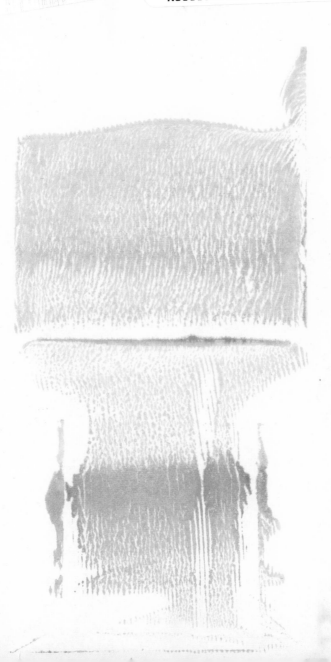

SURFACE WATER SEWERAGE

SURFACE WATER SEWERAGE

RONALD E. BARTLETT

F.I.C.E., F.I.P.H.E., F.I.W.E.S., M.Inst.W.P.C.

Consulting Civil Engineer

A HALSTED PRESS BOOK

JOHN WILEY & SONS
NEW YORK—TORONTO

PUBLISHED IN THE U.S.A. AND CANADA BY
HALSTED PRESS
A DIVISION OF JOHN WILEY & SONS, INC., NEW YORK

Library of Congress Cataloging in Publication Data

Bartlett, Ronald Ernest, 1920-
    Surface water sewerage.

    Bibliography: p.
    Includes index.
    1. Storm sewers. 2. Sewer design. 3. Runoff.
       I. Title.
TD665.B37   1976   628'.21   75-46624
ISBN 0-470-15020-3

WITH 18 ILLUSTRATIONS AND 17 TABLES

© APPLIED SCIENCE PUBLISHERS LTD 1976

Composed by Eta Services (Typesetters) Ltd., Beccles, Suffolk
Printed in Great Britain by Galliard (Printers) Ltd., Great Yarmouth, Norfolk

Preface

As countries have developed there has been a general tendency for people to leave the countryside to live in or near the towns and cities. It was said recently that 45% of the world's population has now become urban. This has resulted in a rapid growth of paved land surfaces so that in many catchment areas there has been a very noticeable increase in the proportion of the area which is relatively impervious. An increase in the paved area of a catchment results in a quicker run-off of the precipitation and also an increase in the *total quantity* of run-off due to the increase in the impermeability of the surfaces. Urban development also usually includes alterations to the drainage patterns, with watercourses being supplemented or replaced by piped sewerage systems.

The design of surface water sewers has developed considerably over the last seventy years or so since the work of Lloyd-Davies and others at the turn of the century. More recently there has been a movement towards the increased use of computers in sewerage calculations. The author has seen numerous examples of overdesign of surface water sewers where the design has been based on rule of thumb methods; this overdesign results in an unnecessary increase in the cost of a scheme. Occasionally schemes are *under*designed where too little allowance has been made for the run-off from areas upstream of any proposed development.

The intention of this book is to cover those aspects of rainfall and run-off which affect surface water sewerage design, together with the hydraulic and structural calculations of the sewers themselves. A chapter on storage has been included as this aspect of surface water run-off has become more important as urban areas have increased.

The metric (S.I.) system has been used throughout the book. A conversion table is included in the Appendices for comparison where necessary.

Following the practice of the author's previous publications, this book is intended mainly for the practising engineer, although it is hoped that students of public health engineering will find it of value. Reference has been made from time to time to the author's earlier book *Sewerage* which contains details of the more general aspects of sewerage work.

In 1969, T. P. Hughes [25] wrote that 'many fields of engineering leave the designer with a finished product on which he and others can gaze with pride—the motorway, the bridge, the reservoir. The usual criterion for the satisfactory sewer, however, is that when complete it shall never be seen or heard of again. The sewerage engineer is seldom congratulated on building a fine structure—but soon hears when the sewage or surface water does not disappear where it should or reappears where it shouldn't.'

I am indebted to members of my staff for their co-operation and assistance in the preparation of the figures included in the book, and for the typing and checking of the manuscript.

Ashby-de-la-Zouch   R.E.B.
Leicestershire

Contents

CHAPTER 1

Introduction

Surface water sewers are provided to carry the run-off from roofs, yards, roads, etc. They are therefore designed according to the extent and type of area to be drained, and must be based on a specific figure for rainfall intensity. If these sewers carry *only* surface water, they can discharge, without treatment, to convenient watercourses. If they form part of a 'combined' system of sewers, they will normally discharge to a sewage treatment works, provision often being made *en route* for storm sewage overflows. These overflows can be sources of pollution of a stream and the modern tendency is therefore towards 'separate' rather than 'combined' sewers.

In temperate climates and in areas of comparatively low rainfall intensities, surface water sewers almost invariably take the form of pipes laid underground; in the tropics, where rainfall intensities are higher, it is more usual (and more economical) to have separate surface water sewers and these are often constructed as rectangular channels, often without top covers.

Surface water sewers need to follow the same basic pattern as a foul sewerage system, i.e. they should be laid in straight lines between manholes, which should be provided at all changes in line and gradient. Surface water sewers should also be laid to *self-cleansing* gradients, as these sewers will carry water laden with grit and other impurities. The basis of design of a surface water sewerage system is however such that the *trunk* sewers are not designed to have a capacity equal to the total capacity of all the branch sewers; this is because the design rate of rainfall at any point in the system is based on a time of concentration, which in turn is dependent on the aggregate length of sewer draining to that point.

1

In recent years there has been considerable research into methods of storage of surface water by its temporary retention in balancing reservoirs, water meadows, etc.; storage results in a decrease in the peak flow downstream, with a subsequent saving in the cost of trunk sewers or in river improvement works.

Surface water sewers provided for highway drainage must be effective not only to clear the road of surface water and prevent it from seeping into the road foundation, but also to prevent mud from the verge, or grit from the road, obscuring the edge line or kerb to the carriageway, and to prevent any adverse effect from existing land drainage.

The legal responsibilities for surface water drainage are complex. Water and local authorities are responsible for the provision of adequate sewerage to effectually drain their districts, and also for the maintenance of watercourses in towns and villages. Highway drainage is the responsibility of the highway authority. Watercourses in open country are the responsibility of the riparian owners and occupiers. 'Main rivers' are the responsibility of the water authorities. It should be noted that only in the case of public sewers does U.K. legislation impose duties on the responsible authorities and that there is no clearcut delimitation of responsibility for flood prevention.

It is often emphasised that a design engineer rarely has all the information he would consider relevant when he is instructed to design a new major surface water sewer for an urban area. Often little or no allowance can be made for future boundary extensions, large-scale development or local government reorganisation; the result is frequently a need for an expensive storm relief sewer at a later date.

The system of drainage to be used for any new development will be dependent on the system of public sewers available off the site. If these sewers are on the combined system, foul sewage and surface water may be discharged into them through a system of combined drains; the surface water may be discharged separately into soak-aways; or a separate system of surface water drains may be constructed to discharge directly to a watercourse. If, however, the public sewers are on a separate system, only the foul sewage from the development may be discharged to the public foul sewer; the surface water may be discharged to the public surface water sewer or independently as before. The majority of local authorities now require that all new development shall be drained on a separate system, whether or not the public sewers are separate, as this will facilitate any separation which may be required at a later date.

EARLY PRACTICE

While it is only comparatively recently that efforts have been made to provide sewers to carry foul sewage, surface water sewerage systems have existed in many parts of the world since ancient times. In the U.K., up to the early part of the 19th century any sewers which existed were for surface water only, and it was illegal to discharge into them what is now termed 'foul sewage'.

The development of the water closet resulted in the construction of cesspools, and gradually it became accepted practice to connect cesspools to the sewers. The rapid increase in the installation of water closets, together with a substantial increase in the discharge of industrial wastes, resulted in many existing sewers becoming used as 'combined sewers' with the subsequent construction of 'sewage farms' and storm sewage overflows. In 1852, when water-borne foul sewerage was in its infancy, it was said that 'the true purpose of town sewers is the instant removal from the vicinity of dwelling houses and from the site of villages, towns and cities of all refuse liable to decomposition and which is capable of being conveyed in water'. By the end of the 19th century it had become an offence to discharge untreated sewage to a river.

'A situation was therefore developing where sewerage systems, many of which had grown from somewhat haphazard surface water drainage arrangements, were required to take greater and greater flows of both foul sewage and surface water as populations and the paving of road surfaces increased, and it is evident that there was a good deal of improvisation to meet changing circumstances. Even today, many towns have a complexity of cross-connections within their sewerage systems that can only have resulted from piecemeal attempts to relieve overloaded sewers by diverting some of the flow into others which at the time had some spare capacity' [10].

Many of the sewerage systems in the world have developed from such beginnings, and although various improvements and separations of flow have been implemented, these old systems still retain many features which severely limit their use; this may become particularly apparent when some major residential or industrial development is planned.

The design of surface water sewers has progressed over the years from earlier practice based on experience only (later checked by measurement), through a stage when the approach was almost entirely theoretical. Later research tends to indicate that the early formulae and rule-of-thumb methods were remarkably accurate, and many engineers are inclined to use the earlier (simpler) methods in preference to later, more complex design methods.

PRELIMINARY INVESTIGATIONS

Surface water sewer design is not an accurate science. It is based on a knowledge of the area to be drained, its gradients and relative impermeability; it must take into account possible and probable future developments; it also takes account of rainfall statistics and then relates these to the cost of sewers so that a balance can be reached between the cost of a scheme and the value of any damage which may result from occasional flooding.

The design of a surface water sewerage system begins with a study of the area to be drained. The limits of that area are usually set by natural physical features, and may extend well beyond the limits of any specific development project. The basic requirement is a map of the area to a scale of not less than 1:2500, and preferably to 1:1000, with contours at least at every metre. Existing natural drainage lines can then be marked on the map and the area being considered can be divided into a series of 'drainage areas'. If an Ordnance Survey map is being used, it will normally be necessary to supplement the information provided on the map with an inspection and survey of the ground.

During the survey, an accurate system of temporary bench marks should be established throughout the area to be covered by the proposed sewers. In most cases, sufficient information should be obtained to enable a map to be prepared with the necessary contours, although in some urban schemes it may be sufficient to obtain levels at street junctions and at other points where the gradients change. Where any existing streams, canals or culverts are encountered during the survey, full details should be recorded of their sizes and levels and of their normal and flood water levels where these can be obtained. In due course it will be necessary to obtain full information on all other services in the vicinity (water, gas, etc.); this can usually be obtained from the relevant authorities.

CLIMATE AND METEOROLOGY

Much of the rainfall recording carried out in the U.K. is based on the use of simple gauges to record the total precipitation during a period of twenty-four hours. This type of information is of relatively little interest as far as sewerage design is concerned, and the installation of a number of autographic recorders since the mid-1950's has helped to provide more useful data on rainfall intensities. The general information available to engineers on rainfall intensities in the U.K. is however normally based on work carried out by Bilham in 1935 [19]; he used data for ten years from eighteen stations and made use of systematic standardised tabulations of rainfall 'duration–intensity' events. His formula is referred to in a later chapter.

Studies carried out subsequently by the Meteorological Office in the U.K. have provided information for the preparation of typical design storm profiles and have tended to show that more storms of high intensity occur over an area than are predicted by the Bilham formula; these storm profiles must, however, be regarded as provisional as they are based on only a few years' data in an isolated area of the U.K. Research is also continuing on the effects of 'dynamic' rainfall, i.e. moving rainstorms that produce non-uniform precipitation in time and space.

Recent research in the U.K. suggests that there is a variation in rainfall intensity over the country, and that in the Midlands and South-East England (where thunderstorms occur more often than elsewhere) certain types of intense rainfall may, in fact, be more frequent. Research in the U.S.A. has suggested that the thermal effects of an urban complex can increase the rainfall in and downwind of the complex by as much as 20%.

GEOLOGICAL AND OTHER FACTORS

The nature of the underground strata of an area will affect to some minor extent the percentage of the rainfall which will result in run-off. The major effect on the amount of run-off is the relative impermeability of the *surface*.

Where soakaways are proposed, the relative permeability of the substrata then becomes important. In many localities the use of soakaways would reduce the cost of surface water sewerage; in some cases they may be technically *preferable* to a piped system of sewers. Where there have been no natural watercourses before development is proposed it may usually be assumed that the subsoil is suitable for soakaways and that any other form of artificial surface water sewerage system could result in an overloading of the existing streams and rivers.

A satisfactory soakaway system requires not only a suitable subsoil such as gravel or fissured chalk, but also a water table low enough to allow a free flow of the surface water into the subsoil at all times of the year.

CHAPTER 2

Rainfall

SOURCES OF PRECIPITATION

Precipitation of moisture results from the condensation of water vapour in the atmosphere. This is often associated with fronts and 'frontal' storms. A frontal system, moving with a velocity of up to 30 km/h, will produce a wide belt of rainfall over a prolonged period. Within this belt there will be pockets of higher intensity precipitation which may move very rapidly. Other storms may be 'orographic' (due to air masses being forced upwards over hills and mountains), or they may be caused by 'thermal convection' due to local differences in temperature.

As the warm air of a 'warm front' rises over a colder air mass, it cools and clouds are formed. Stratus-type clouds are rain-bearing clouds and under these conditions heavy rain is likely for about 200 to 300 km ahead of the front itself. As the front passes, the rain will cease and the temperature will rise. With a cold front, the warm air is forced rapidly upwards by the moving cold air mass, this becomes very unstable and may result in the formation of cumulus-type clouds, with associated thunderstorms. A belt of rain, often 80 or 90 km across, will usually accompany a cold front.

Thermal convection storms normally take the form of thunderstorms or 'cloudbursts'. Their extent is normally fairly limited, and for comparatively small catchments (below about 2500 km^2) the run-off from this type of storm may be greater than from a frontal storm.

Clouds are produced by the condensation of water vapour due to a fall in temperature of the air. The small drops of water in a cloud (average diameter about 40 μm) tend to fall slowly towards the earth but are so light that they may sometimes be carried upwards by convection currents. If the drops continue to cool, more water

7

vapour may condense on them so that eventually they are too heavy to be supported by the ascending air and they fall as raindrops. Rain is often the result of the fine drops coalescing around particles of dust, salt, ice, etc. to form into raindrops. This is the principle behind the 'seeding' of clouds with dry ice (frozen CO_2), or with silver iodide, to produce rainfall.

While 'precipitation' in the present context includes snow, sleet, hail, dew and frost, the more usual precipitation taken into account when considering sewerage problems is that of rain (droplets of 0·5 mm and over) or drizzle (droplets less than 0·5 mm). Drizzle usually originates from low stratus cloud or from fog. Data issued by most meteorological departments when referring to 'rainfall' intend the expression to cover all forms of precipitation expressed as equivalent rainfall.

Snowfall is usually measured according to its undisturbed depth. This can be converted to an equivalent amount of rainfall by taking 1 m depth of snow as equivalent to about 80 mm of rainfall. Snow may be of particular importance, either because it can form a substantial part of the annual precipitation in some areas, or because of its effect on the *rate* of run-off, either when it covers the ground or when the snow begins to melt. The rate of run-off can be predicted with reasonable accuracy, based on knowledge of the processes of storage, evaporation and melting [30].

STATISTICS

Statistics of annual and maximum rates of rainfall are prepared by most countries. The first known rainfall records in the U.K. were made at Townley near Burnley in the 17th century. Rainfall data are now published annually in *British Rainfall* in the U.K.; by the U.S. Weather Bureau; and by the meteorological departments in many other countries. Published information, however, rarely includes details of rainfall *intensities* which are so important in the design of surface water sewerage systems. Intensity is expressed in terms of millimetres per hour and is related to a statistical frequency, e.g. once per year, once in 10 years, etc.

Methods of rainfall measurement in the U.K. have been standardised and are summarised in *Rules for Rainfall Observers* [8], which sets out standards for the location of gauges and for their reading. These rules must be observed if accurate results are to be obtained, and if results are to be comparable with those of other rainfall stations. Standards for siting of rainfall gauges and the errors which might occur were discussed by Rodda in a paper in 1967 [31].

Three types of gauge are fairly standard throughout the world. The most common is the simple non-recording gauge which records the amount of rainfall over some predetermined period of time (often 24 h). In the U.K., these gauges are usually 127 mm diameter, placed so that their rims are 305 mm above the surrounding ground level; measurements are made at 0900 hours each day. A few gauges used in the U.K. are 203 mm diameter. The U.S. Weather Bureau uses gauges of 203 mm diameter (32 400 mm^2); the French Association uses collectors of 40 000 mm^2; while some other European countries have adopted the Hellman gauge with a collector area of 50 000 mm^2. Two types of gauge suitable for recording the *rate* of rainfall use either a weighing device or a tipping bucket; some countries use a float type with a siphon emptying arrangement. Recording charts are provided with these gauges and a careful analysis of the charts over a period of time will give a fairly accurate indication of rainfall intensities for various durations of storm. Special rain gauges are available to measure rainfall *intensities* but these are quite rare.

There are now about 5000 rainfall stations forming an irregular distribution over the U.K.; their locations generally correspond with the distribution of population, with additional gauges at the sites of reservoirs for water supply and hydro-electric power. The location of gauges is governed, to some extent, by economics; for long-term average figures over large areas relatively few gauges are needed, but they must be more closely sited to record intense local storms. As a guide, for an area of 1000 ha, two or three gauges may be required for adequate coverage; when the area is (say) 20 000 ha, seven or eight gauges may be necessary.

Limited knowledge of rainfall intensities is always better than none, and even a few years of readings from a recording gauge can provide a useful guide in surface water sewer design. The first recording gauge to be installed in Sri Lanka (Ceylon) was at Colombo in 1898. The following year the Surveyor-General reported that 'the table showing the intensity of rainfall at Colombo has already been found valuable in connection with the drainage system of the town'. As that city's drainage system appears to have given little trouble during its 50 or 60 years of life, it would appear that the intelligent use of even one year's records can be worthwhile.

AVERAGE RAINFALL

Almost every country has published records of average rainfalls. These will vary across a country depending on the proximity of oceans and on the contours of the land. The average rainfall is usually higher in the vicinity of high mountain ranges—this is very noticeable in Scotland and in North-West England, and along the western seaboard of the U.S.A. The average annual precipitation over the whole of the U.K. is of the order of 1100 mm, while in the mountainous regions of the North-West this average can be as high as 5000 mm. The mean value of rainfall over any particular area is often determined by the Theissen polygon method [16].

Over the last 100 years a daily rainfall of about 160 mm or more has occurred about once every 4 years somewhere in the U.K.; falls of this order are most frequent in the South-West, and are least frequent in the Midlands and the South-East. On the other hand, it appears that some forms of intense rainfall are *more* frequent in the Midlands and South-East; this may be due to the greater frequency of thunderstorms in these areas. An investigation during storm rainfalls in South Lancashire during 1960 to 1964 showed that rainfall intensities could vary from 1·0 mm/h for a storm lasting 10 h to 6·5 mm/h for a storm of 6 h. The maximum daily rainfall recorded anywhere in the U.K. is about 280 mm.

It is generally accepted that there is an upper limit to the amount of rainfall likely to fall over a specified area in a given time; this is referred to as the Probable Maximum Precipitation. This was discussed by Wiesner in a paper to the Institution of Civil Engineers in 1964 [33].

RAINFALL INTENSITIES AND FREQUENCIES

Average rainfall has, however, only a limited bearing on the design of surface water sewerage schemes. The basic principles of surface water sewerage design are based on the *intensities* of rainfall over given periods of time. It will often be found that rainfall intensities, in terms of rainfall in periods of less than about 1 h, are similar throughout a small country irrespective of ground contours. In larger countries, it will be necessary to consider variations between one part of a country and another.

The rate of rainfall to be used in any sewerage calculations must be based on a chosen duration of storm. The relevant period of time will depend on the 'time of concentration' of the particular drainage area being considered, and will therefore vary with each section of the calculations (see Chapter 3). The mean rate of rainfall must then be determined for the duration of storm according to a given frequency of recurrence. Intensity expressed in terms of millimetres per hour is therefore related to a frequency, e.g. once per year.

Rainfall records throughout the world have been analysed and tabulated so that a rate of rainfall can be determined statistically according to storm frequency and duration. Data for storms in the U.S.A. have been recorded by the Corps of Engineers, storms in the U.K. are referred to in *British Rainfall*, while data for various other countries are published in their meteorological department records. In many parts of the world, this information has been summarised in the form of one or more formulae, so that the choice is then a fairly simple matter of choosing a duration of storm to suit design considerations and a frequency to suit the economics of the proposals.

When formulae have not been prepared, estimates of rainfall intensities for varying lengths of storm can be obtained by abstracting details from the charts of rainfall recorders where these are available. Reasonably accurate formulae for '1-year' storms can, for example, be obtained by plotting the amount of rainfall during each storm of a certain duration against the number of occurrences of such a storm during the year's records. If this is plotted on a log–log graph paper it is then possible to interpolate '1-year' storm intensities for each period of time.

Codes of Practice in use in Europe and the U.S.A. rely on the use of different formulae for different regions of a country. For example, the basic formula used in the U.S.A. takes the general form of

$$I = \frac{KN^x}{a+t^n} \qquad \qquad \textbf{Formula 2.1}$$

where

I is the average rainfall intensity,
N is the frequency in years that the intensity is reached,
t is the duration of the rainfall, in min,
K, x, a and n vary with geographic location.

In the U.K., reliance is usually placed on the 'Ministry of Health' formulae, or the Bilham or Norris formulae. The Ministry formulae take the form

$$R = \frac{a}{t+b}$$

where

    R is the rate of rainfall,
    t is the duration of the storm,
    a and b are constants.

The original Ministry formulae, expressed approximately in metric units, were:

  (i) for $t = 5$ to 20 min;

$$R = \frac{750}{t+10} \text{ mm/h}$$
                                               Formula 2.2

  (ii) for $t = 20$ to 120 min;

$$R = \frac{1000}{t+20} \text{ mm/h}$$
                                               Formula 2.3

where

    R is the rate of rainfall, in mm/h,
    t is the duration of the storm, in min,
    (i.e. the time of concentration plus the time of entry).

The two Ministry formulae relate to storms of a frequency expected about once per year. In 1935, Bilham published his formula, which relates the frequency of storms to their duration and intensity. Expressed in metric terms, Bilham's formula is

$$T = 1\cdot25t\,(0\cdot0394r+0\cdot1)^{-3\cdot55}$$
                                               Formula 2.4

where

    t is the duration of the storm, in h,
    r is the total amount of rainfall during time t, in mm,
    T is the number of storms of this intensity to be expected in 10 years.

Rainfall intensities according to the Bilham formula are illustrated graphically in Fig. 2.1. Simple 'Ministry-type' formulae, based on work originally published by Norris in 1948, are set out in Table 2.1.

Formula 2.3 is in fact the formula suggested by Lloyd-Davies in 1906 [26], and often referred to as the 'Birmingham Curve'; Formula 2.2 was put forward by a Ministry of Health Committee in 1930 as a modification of the Birmingham Curve for storms of shorter duration.

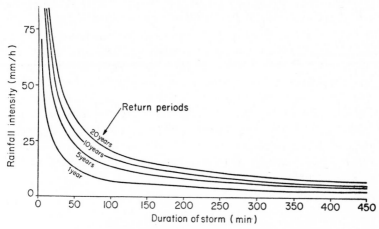

Fig. 2.1. Rainfall intensities based on Bilham's formula.

TABLE 2.1

RAINFALL INTENSITIES

Frequency of recurrence, once in	Intensity (mm/h)	
	t = 5 to 20 min	t = 20 to 120 min
0·5 year	$\dfrac{580}{t+10}$	$\dfrac{760}{t+19}$
1·0 year	$\dfrac{660}{t+8}$	$\dfrac{1000}{t+20}$
2·0 years	$\dfrac{840}{t+8}$	$\dfrac{1200}{t+19}$
5·0 years	$\dfrac{1220}{t+10}$	$\dfrac{1520}{t+18}$
10·0 years	$\dfrac{1570}{t+12}$	$\dfrac{2000}{t+22}$

Research carried out by the Meteorological Office has suggested that standard storm profiles can be drawn relating the rate of rainfall with the time elapsed since the start of a storm. Separate profiles have been drawn for different frequencies of storm. These are illustrated in Road Note No. 35 [13]. These were based on studies using autographic rain recorders and investigated rainfall profiles and the design depth for an areal fall for a given frequency and duration; the studies covered areas of the order of 1000 ha. It has been suggested that when it is necessary to consider rain falling over a larger area, consideration should be given to long duration heavy falls of rain recurring more frequently than are predicted by the Bilham formula.

CHAPTER 3

The Design Storm

In the U.K., intense rainfalls during short periods of time occur most commonly with thunderstorms or with showers of a thundery type. These intense falls are not related in any way to the distribution of average annual rainfall, and their frequency is usually at a maximum during the summer months.

Early calculations for sewerage were often based on flat rates of rainfall irrespective of the area of the catchment. In 1935, it was proposed that the intensity of rainfall should be varied to suit the size of the catchment area, and that the figures to be used should vary from 6 to 25 mm/h, depending on the area of the district, the smaller figures being used for larger sewers. The use of a flat-rate rainfall intensity is, in fact, still normal practice for very small areas, i.e. less than about 0·25 ha, or where the 'time of concentration' (see below) is about 15 min or less.

For years many architects have used a flat-rate figure of 25 mm/h for the rainfall when calculating the run-off from industrial and housing developments, irrespective of the area involved; in more recent years figures varying from 12 to 25 mm/h have been used by some designers to calculate run-off from impervious areas. The Code of Practice on Housing Drainage states, however, that it is desirable for 'domestic drainage work' in the U.K. to be designed on the basis of a rainfall intensity of 38 mm/h, while the Building Research Station [2] has recommended that the drainage in the immediate vicinity of buildings should generally be designed on a rainfall intensity of 50 mm/h. When the Lloyd-Davies formula (see Chapter 4) was first published, many engineers felt that rainfall intensities of 50 mm/h and more were unreasonable; the Code on Sewerage [1]

15

proposes a 'ceiling' figure of between 25 and 38 mm/h when using traditional methods of calculating sewer sizes. Lower figures will then be used when these are the result of calculations using one of the accepted formulae.

As stated in Chapter 2, the choice of storm intensity is based partly on the duration of a storm and partly on the statistical frequency of recurrence to be used. In his paper in 1906 [26], Lloyd-Davies referred to work carried out in the U.S.A. by a Mr. Emil Kuichling and he then illustrated that when rainfall run-off records were available it could be shown that for storms of a selected frequency of recurrence the one which produced the highest run-off would be the storm whose duration equalled the *time of concentration* of the catchment being considered. If the length of storm is taken as *greater* than the time of concentration, the rate of rainfall will be less. If, however, the length of storm is taken as *less*, although the rainfall intensity will be greater, the rain falling on the upper part of the catchment will only reach the lower part at the end of the *time of concentration*, by which time the precipitation will have ceased in the lower section.

The 'rational' method of approach, used by Lloyd-Davies, makes the assumption that the whole of the catchment is contributing to the point under consideration at a time after the start of the rainfall equal to the time of concentration for the area and that the peak rate of run-off occurs at the same time. This time of concentration is the time taken for run-off to flow to the point under consideration from the furthest point in the system, taken at full-bore velocity, and with an allowance for a 'time of entry'. This can be expressed as

$$t_c = \frac{\text{length of sewer}}{\text{full-bore velocity of flow}} + \text{time of entry}$$

It is recommended that a time of entry of 2 min should be used for normal urban areas, increasing to up to 4 min for areas with exceptionally large paved areas with slack gradients. A time of entry should not be estimated to an accuracy greater than about 30 s. The design storm is taken as that storm having a duration equal to the time of concentration.

It should be remembered that the time of concentration is used in both the rational (Lloyd-Davies) method of calculating the run-off and in the T.R.L.L. hydrograph method. This latter method is based (*inter alia*) on the proposition that the effect of the variation in the rate of rainfall during a storm may be determined by the use of a storm profile which is a statistical average of recorded storms.

When deciding on a frequency of rainfall recurrence, allowance must be made for property values and for the probable value of any damage which might be caused by occasional flooding; a compromise is then reached and a 'once-per-year', 'once-in-2-years', or 'once-in-5-years' storm will be chosen. General practice in the U.K. is to use storm frequencies of once-per-year or once-in-2-years for most sewerage schemes, with once-in-5-years storms being adopted where property in vulnerable areas would be subject to appreciable damage if flooding should occur; frequencies of 10 or 25 years may be adopted for city centre sewers. Practice in the U.S.A. tends towards 'once-in-5-years' for sewers in residential areas and between 10 and 50 years for high-value districts. There is of course considerable difference between foul sewage flooding in houses and storm run-off flooding on to open land; the *type* of flooding which is likely to occur must always be considered. Highway drainage schemes are usually designed on a '1-year' storm basis; higher rainfall intensities which occur from time to time will then result in water standing on the highway for short periods of time.

Flood protection schemes for river valleys may also be based on figures of 10 to 25 years. While river channel improvements in open country may be designed on 5-year storms, it would be usual to design improvements in urban areas on recurrences of 50 to 100 years. Culverts and small bridges under main highways are generally designed for a 50- or 100-year recurrence; on less important roads 25 years may be adopted.

While it may sometimes be suggested that storms of (say) 50 or 100 years frequency should be used in calculations, these would result in extremely expensive sewers, and investigations by the U.K. Meteorological Office have shown that storms of this magnitude rarely cover the whole of the catchment area of a sewer. The actual run-off from this magnitude of storm affecting only a part of a catchment could be less than that of a smaller storm spread over the whole catchment.

In all references to average storm frequencies, it must be remembered that a '5-year' storm does not mean that the frequency can be expected to occur at 5-year intervals, but rather that over a longer period of (say) 50 years, this intensity of storm will statistically occur 10 times; there could be two or more occurrences in a single year.

Research is now being carried out in the U.K. to investigate the relationship between storm frequency, duration and mean rate of rainfall in different parts of the country. Until these investigations have been completed, it is advisable to analyse local data where records are available. For general use and where local records are not available, the Bilham or the Norris formulae are recommended; where a '1-year' storm is to be used the 'Ministry' formulae are available.

Sewers are almost invariably laid at least a metre below the surface and they can therefore accommodate a fairly considerable surcharge before the point is reached where flooding is almost occurring. The discharge capacity of the sewers under these surcharge conditions is much in excess of their theoretical capacity—this may be as much as doubled. Inspection of Fig. 2.1 will show that a 10-year storm will give a rainfall figure roughly twice that of a once-per-year storm for durations up to about 50 min. It follows that where sewers have been designed on a once-per-year storm capacity, surcharge can probably increase that capacity up to the equivalent of a 10-year storm without any damage from flooding.

CHAPTER 4

Run-off

In very general terms, run-off is that part of the rainfall (or precipitation) that drains from the land after allowance has been made for evaporation. The *rate* of run-off is affected by retention on the surfaces, by infiltration into the surface and by the hydraulic characteristics of channels and gullies used to carry the flow into the sewers. While it would be possible to produce equations for each of these factors, they could not be solved in practice, and the study of run-off is therefore often based on actual measurements of flow, or simplified approximations of the various conditions have to be made.

IMPERMEABILITY

Vegetation will intercept a significant part of rainfall in rural areas, and for any storm this will range from 0·25 to 12·0 mm according to the vegetation. Further significant retention (as far as sewer design is concerned) occurs at pervious surfaces such as lawns and in surface depressions which are present in virtually all surfaces (this can amount to up to 2·5 mm for any storm). This aspect of surface water sewer calculations is more generally referred to as the 'impermeability' or the *impermeability factor* for the catchment; in some countries this may be referred to as the 'run-off factor'.

It will be apparent that under normal circumstances only a comparatively small part of the rain falling on to a surface will find its way into the sewers; the remainder will percolate into the ground, will be evaporated or will be held up (permanently or temporarily) by ponding on surfaces and in gutters, etc. Even apparently impervious surfaces are pervious to some extent. It is therefore rare to assume that the run-off will equal 100% of the rainfall. Many tables of impermeability factors have been published based on experience and research; the rate of rainfall is then multiplied by the relevant impermeability factor when calculating the rate of run-off. Table 4.1 has been compiled from a number of sources.

TABLE 4.1

IMPERMEABILITY FACTORS

Type of surface	Factor (%)	Type of surface	Factor (%)
Urban areas, where the paved areas are considerable	100	heavy clay soils	70
		average soils	50
Other urban areas, average	50–70	light sandy soils	40
residential	30–60	vegetation	40
industrial	50–90	steep slopes	100
playgrounds, parks, etc.	10–35	Housing development at	
General development—		10 houses per hectare	18–20
paved areas	100	20 houses per hectare	25–30
roofs	75–95	30 houses per hectare	33–45
lawns—depending on slope and subsoil	5–35	50 houses per hectare	50–70

In the past, attempts have been made to distinguish between 'ultimate' impermeability and 'effective' impermeability. The former relates to the difference between rainfall and run-off over a long period of time; the latter relates to the impermeability at the end of the relevant time of concentration. In his original paper [26] Lloyd-Davies referred to the change of impermeability with time, and stated that 'the impermeable percentage gradually increases with the duration of rainfall'. A formula published by Escritt, based on research, gave the effective impermeability at any moment in time as

$$P = i \left(1 - \frac{2}{t}\right)$$ Formula 4.1

where

    P is the effective impermeability,
    i is the ultimate impermeability determined from total rainfall and total run-off,
    t is the time after the start of the rainfall, in min.

Except for short storms, of less than (say) 15 min duration, it is usual to design sewers on the *ultimate* impermeability.

While the use of an impermeability factor is normal, it should always be remembered that these are generally *maximum* figures, and that frequently the run-off from any particular surface will be lower than indicated. This is because the rate of run-off will also depend on previous conditions before the storm; the rate of run-off from a dry catchment is much lower than from saturated or frozen ground.

The choice of impermeability factor will usually depend on the overall characteristics of the area being drained. For an industrial development, it may be more convenient to measure the totals of each type of surface (roofs, roads, grassed areas, etc.) and to multiply each by a suitable factor. For housing development, on the other hand, it is often more satisfactory to adopt a factor according to the proposed housing density and to multiply the *whole* of the area by that factor. Escritt [40] has suggested that, in some circumstances, all paved and roofed areas could be assumed to be 80% impermeable, and all unpaved earth surfaces to be 100% permeable. As an approximate guide to rates of run-off, a rainfall intensity of 50 mm/h, on a 100% impermeable surface, will produce a run-off of $0.14 \text{ m}^3/\text{s/ha}$ (0.14 cumec/ha).

The impermeability factors quoted in Table 4.1 are suitable for use with the 'rational' (Lloyd-Davies) method of calculation. The U.K. Department of Scientific and Industrial Research (in Road Research Technical Paper No. 55 [12]) proposed that when using the *unit hydrograph* method, and 'subject to some qualifications in exceptional circumstances', the whole area of paved surfaces in an urban area should be considered impermeable in sewer design calculations, and the *unpaved* areas should be considered completely pervious.

It has been pointed out [13] that under British rainfall conditions, in the majority of *urban* areas, the total surface area contributing to the flow in surface water sewers can usually be taken as only the area of paved surfaces directly connected to the sewer system. This would include footpaths from which the surface run-off passes to a gully, but would not include the run-off from any paths which drain to unpaved areas. In some special cases the 'impermeable area' should include cuttings; some additional allowance may also be required for *large unpaved* areas which will drain to a sewer.

THE RATIONAL (OR LLOYD-DAVIES) METHOD

In 1906, when Lloyd-Davies gave his paper to the U.K. Institution of Civil Engineers [26], he showed that the storm to produce the greatest run-off was that with a duration equal to the time of concentration of the catchment area being considered. This method appears to have been used in Ireland in 1851 and in the U.S.A. from 1889. The general expression of the Lloyd-Davies method, and of the American 'rational' method, is set out in the formula

$$q = ciA$$

where

> q is the peak rate of flow,
> c is a constant based on the impermeability of the catchment,
> i is the mean rate of rainfall during the time of concentration,
> A is the catchment area.

In finite terms the above formula becomes

$$Q = 2 \cdot 75\, ApR \times 10^{-3} \qquad \textbf{Formula 4.2}$$

where

> Q is the run-off, in cumec,
> Ap is the impermeable area, in ha,
> R is the rate of rainfall, in mm/h.

In this formula, the impermeable area (Ap) is the 'equivalent impervious area' and is the total area of the catchment (in hectares) multiplied by a suitable impermeability factor. The choice of factor should always be based on what is to be expected in the foreseeable future, particularly where development is intended. As points more remote from the head of the sewer are considered, the values of Ap and R must be adjusted accordingly.

The Lloyd-Davies formula is a mathematically accurate formula, but its very simplicity is based on certain assumptions which in themselves may contain errors. These assumptions are that:

1. the rate of rainfall is constant throughout the storm,
2. the impermeability of the catchment area remains constant,
3. the velocity of flow in the sewers remains constant at full velocity throughout the time of concentration,
4. the time–area graph is a straight line, i.e. the impermeable area is evenly distributed over the catchment.

It is obvious that none of the above four assumptions will normally apply in practice, and while the errors resulting from the assumptions to some extent will balance themselves out (items '1' and '4' tend to underestimation, while '2' tends to overestimation), it is generally accepted that the Lloyd-Davies method overestimates the magnitude of the run-off.

The method is, however, frequently used because of its simplicity. It is used in the U.S.A. and was recommended by the U.K. Department of Scientific and Industrial Research [12, 13] where the diameter of the largest sewer is unlikely to exceed 600 mm, i.e. for 'small areas'. The rational method is used extensively in the U.S.A. for catchments up to about 12 km^2. For larger areas it is generally felt that development of data for application by hydrograph methods is warranted. Details of the procedure for using the rational method are set out in Road Note No. 35 [13]. When understood and when used correctly, the method generally gives satisfactory results.

Use of the method entails the definition of the drainage tributary to any point under consideration, and the preparation of a preliminary layout of sewers, with tentative locations of inlet points. The procedure is (to some extent) one of trial and error, as the first calculations must be based on an assumed sewer diameter and gradient to give the first assumption for time of concentration. The calculation must therefore be repeated as necessary until the calculated sewer diameter equals that assumed when choosing the time of concentration.

THE TIME–AREA GRAPH

Among the modifications proposed for the rational method were various more complicated methods involving time–area graphs. The use of a time–area graph aimed at making allowance for the irregular disposition of the impermeable areas within the catchment, i.e. the elimination of the error under item '4' above. However, as stated earlier, the errors under '4' of the rational method were, to some extent, balanced by overestimation under item '2'. Users of the time–area graph method therefore, while aiming at greater accuracy, in fact produced larger errors of overestimation of the run-off.

One form of time–area method which was used in the U.K. was described by Ormsby and Hart [28]. Its decline in use is due mainly to its tendency to overestimate the rate of run-off.

THE TANGENT METHOD

The tangent method was first published by Reid in 1927 [29] for use with the U.K. 'Ministry of Health' rainfall formulae. It was basically a form of time–area graph, but was less complex than that proposed later by Ormsby and Hart. Modified tangent methods have subsequently been put forward, including a method proposed by Escritt in 1946 [23].

THE UNIT HYDROGRAPH

The general theory of the unit hydrograph is that at a given point on a stream or sewer the base of the hydrograph of the direct run-off from a storm of 'unit duration' is constant, regardless of the volume of run-off, while the ordinates of the hydrograph vary directly as the run-off. A *unit hydrograph* refers to 'unit run-off'. A unit hydrograph can therefore be the discharge hydrograph resulting from 'effective' rainfall falling uniformly over the area, at a uniform rate, and in a specified unit period of time. The unit of time is arbitrary except that it must be less than the time of concentration.

'The practical application of the unit hydrograph method to the calculation of an actual run-off hydrograph involves the selection of a unit hydrograph appropriate to the area, from data recorded at the area or at a similar area. The selected hydrograph is divided into a number of equal time intervals and the area contained between successive time intervals is expressed as a percentage of the total area of the hydrograph. This percentage is known as a distribution factor. The run-off hydrograph for any given rainstorm is calculated from the rainfall curve by multiplying, for each individual time interval, the effective rate of rainfall by the distribution factor and by the impermeable area' [12]. The hydrograph method was developed to overcome errors in the rational method of design due to the assumptions that the rate of rainfall is constant and that the distribution of areas within the catchment is uniform.

In the U.K., a hydrograph method, known as the 'T.R.R.L. Hydrograph Method' has been adopted to provide an accurate and reliable sewer design for any urban area; it is particularly applicable to larger areas. The procedure for using this method is included in Road Note No. 35 [13]. Although the data required for the method are no more than are required for the 'rational' formula, the calculations can normally be carried out only by an electronic digital computer. 'The operations carried out at each manhole by the computer are broadly to calculate a storm profile corresponding to the time of concentration, to calculate a hydrograph of flow for an assumed pipe size from this profile and the area/time diagram, to modify this hydrograph to allow for variations in retention in the system, and to repeat the whole calculation as necessary until the assumed pipe size is large enough to carry the computed peak rate of flow. The computer provides the diameter of each pipe, the maximum rate of run-off in each pipe, and the maximum rate of flow that each pipe can carry' [13].

To some extent, the T.R.R.L. Method is a development of the earlier Ormsby and Hart method. It was devised to overcome the deficiencies of earlier methods and is broadly based on the principles of the unit hydrograph method. It is based on the propositions that the variation in rate of rainfall during a storm may be determined by the use of a storm profile which is statistically an average of recorded storms, and that the effect of the variation in retention of flow in a sewer may be calculated by considering it to be a single reservoir upstream of the point under consideration.

The hydrograph represents the rise and fall of the rate of flow in the sewer during the storm, the peak of the hydrograph being taken for sewer design purposes. An area–time diagram is calculated, and using a suitable design storm profile, the inflow curve hydrograph is prepared. After allowing for retention in the sewers, a second hydrograph (outflow curve) is then prepared.

While the T.R.R.L. Method has been accepted as the leading design method for larger surface water sewerage systems in the U.K. since 1963 and has been the subject of favourable comment by American researchers, it has been criticised by some engineers who consider that for certain categories of area–time diagrams it over-estimates the run-off. It has been suggested that this is due to the rigid application of fixed values for time of entry and for percentage run-off throughout the duration of the storm, and the retention correction to route floods through the sewers. Many engineers feel that the rational method, if intelligently applied, can yield results as good as those obtained by the more complex T.R.R.L. Method and without the need of access to expensive computer time.

PROVISION FOR THE FUTURE

Whichever method is adopted for the calculation of run-off, consideration must always be given to the probable future conditions within a drainage area. Structure Plans, Development Plans, and other planning proposals must be carefully considered, and allowance must be made for any probable development within the catchment in the foreseeable future. This development will normally have the effect of increasing the impermeability factor (and therefore the rate of run-off) as it will entail a decrease in agricultural land and an increase in the area of roofs, roads and other relatively impervious areas.

CHAPTER 5

Hydraulic Design

The calculation of the maximum flow to be accommodated in a surface water sewer will be based on the rainfall and run-off information considered earlier (Chapters 2, 3 and 4). However, while in foul sewer design the designer can choose a pipeline to suit a figure which represents the daily flow rate and therefore to ensure a self-cleansing velocity at least once per day, this is not possible for surface water sewer design. The daily flow in these sewers may be little or nothing for long periods of time. The organic load in a separate surface water sewer should, of course, be very small, but the grit loading may be substantial, particularly at the commencement of a storm when the first flush of impurities is washed off roads and paths into the sewers.

It is, therefore, usual to design surface water sewers so that they have a 'self-cleansing velocity' when running full (and therefore also when not less than half-full—see Table 5.1). A self-cleansing velocity for foul sewers is normally taken as between 0·75 and 1·00 m/s; for surface water sewers this figure *could* be reduced to about 0·6 m/s for 'normal' flows (i.e. when flowing at considerably less than full capacity). As these sewers rarely run more than about a quarter full, it is generally more usual to use the same velocity of 0·75 to 1·00 m/s for the design velocity for a surface water sewer *when running full*.

DRAFT SEWER LAYOUT
Before the sewerage design is commenced, information will be available on a proposed housing or factory layout, or on the general proposals for larger development. With this information plotted on a contour plan to 1:500 or 1:2500 (or even 1:10 000 if relevant), draft proposals for the main surface water sewers can then be set out. The possible points of discharge (e.g. watercourses) will set the overall

27

TABLE 5.1

PROPORTIONATE VALUES OF VELOCITY AND DISCHARGE FOR PIPES RUNNING
PARTLY FULL

Prop. depth	Prop. velocity	Prop. discharge	Prop. depth	Prop. velocity	Prop. discharge
0·01	0·089 0	0·000 2	0·41	0·913 2	0·352 5
0·02	0·140 8	0·000 7	0·42	0·923 9	0·368 2
0·03	0·183 9	0·001 6	0·43	0·934 3	0·384 1
0·04	0·222 1	0·003 0	0·44	0·944 5	0·400 3
0·05	0·256 9	0·004 8	0·45	0·954 4	0·416 5
0·06	0·289 2	0·007 1	0·46	0·964 0	0·433 0
0·07	0·319 4	0·009 8	0·47	0·973 4	0·449 5
0·08	0·348 1	0·013 0	0·48	0·982 5	0·466 2
0·09	0·375 2	0·016 7	0·49	0·991 4	0·483 1
0·10	0·401 2	0·020 9	0·50	1·000 0	0·500 0
0·11	0·426 0	0·025 5	0·51	1·008 4	0·517 0
0·12	0·450 0	0·030 6	0·52	1·016 5	0·534 0
0·13	0·473 0	0·036 1	0·53	1·024 3	0·551 3
0·14	0·495 3	0·042 1	0·54	1·031 9	0·568 5
0·15	0·516 8	0·048 6	0·55	1·039 3	0·585 7
0·16	0·537 6	0·055 5	0·56	1·046 4	0·603 0
0·17	0·557 8	0·062 9	0·57	1·053 3	0·620 2
0·18	0·577 5	0·070 7	0·58	1·059 9	0·637 4
0·19	0·596 5	0·078 9	0·59	1·066 3	0·654 6
0·20	0·615 1	0·087 6	0·60	1·072 4	0·671 8
0·21	0·633 1	0·096 6	0·61	1·078 3	0·688 9
0·22	0·650 7	0·106 2	0·62	1·083 9	0·706 0
0·23	0·667 8	0·116 0	0·63	1·089 3	0·722 9
0·24	0·684 4	0·126 3	0·64	1·094 4	0·739 7
0·25	0·700 7	0·137 0	0·65	1·099 3	0·756 4
0·26	0·716 5	0·148 0	0·66	1·103 9	0·773 0
0·27	0·732 0	0·159 4	0·67	1·108 3	0·789 3
0·28	0·747 0	0·171 2	0·68	1·112 4	0·805 5
0·29	0·761 8	0·183 4	0·69	1·116 2	0·821 5
0·30	0·776 1	0·195 8	0·70	1·119 8	0·837 2
0·31	0·790 1	0·208 6	0·71	1·123 1	0·852 7
0·32	0·803 8	0·221 7	0·72	1·126 1	0·868 0
0·33	0·817 2	0·235 2	0·73	1·128 8	0·882 9
0·34	0·830 2	0·248 9	0·74	1·131 3	0·897 6
0·35	0·843 0	0·262 9	0·75	1·133 5	0·911 9
0·36	0·855 4	0·277 2	0·76	1·135 3	0·925 8
0·37	0·867 5	0·291 8	0·77	1·136 9	0·939 4
0·38	0·879 4	0·306 6	0·78	1·138 2	0·952 4
0·39	0·890 9	0·321 7	0·79	1·139 1	0·965 2
0·40	0·902 2	0·337 0	0·80	1·139 7	0·977 5

TABLE 5.1—*continued*

Prop. depth	Prop. velocity	Prop. discharge	Prop. depth	Prop. velocity	Prop. discharge
0·81	1·140 0	0·989 2	0·91	1·120 0	1·070 1
0·82	1·139 9	1·000 4	0·92	1·115 0	1·073 2
0·83	1·139 5	1·011 0	0·93	1·109 3	1·075 2
0·84	1·138 7	1·021 1	0·94	1·102 7	1·075 7
0·85	1·137 4	1·030 4	0·95	1·095 0	1·074 5
0·86	1·135 8	1·039 1	0·96	1·085 9	1·071 4
0·87	1·133 7	1·047 1	0·97	1·075 1	1·065 7
0·88	1·131 1	1·054 2	0·98	1·061 8	1·056 7
0·89	1·128 0	1·060 5	0·99	1·043 7	1·041 9
0·90	1·124 3	1·065 8	1·00	1·000 0	1·000 0

pattern, and to keep costs to a minimum the sewers will generally follow the contours of the original ground; this will reduce the depths of excavation and should avoid any necessity for pumping in the ultimate layout.

From this draft plan it is now possible to plot sections along the proposed sewers. Once any revisions have been incorporated to avoid obstacles such as existing or proposed underpasses, foul sewers and other services, a schedule of probable sewer sizes and gradients can be prepared.

FRICTION
Since the publication of a number of empirical formulae late in the 19th century, many tables of proposed friction coefficients have been prepared; some of these are probably more relevant in connection with the design of larger diameter pipelines and channels, particularly those constructed to carry clean water.

While many new materials have been developed for the manufacture of pipes, and some manufacturers have put forward claims for low friction coefficients, it needs to be borne in mind that the flow capacity of a sewer after a short period of use may depend upon the characteristics of the slime growing or deposited on the pipe wall and upon the deposit of grit on the invert. 'For this reason it is no longer regarded as necessarily axiomatic that sewers of up to 900 mm diameter be designed for hydraulic roughness dependent only upon the material of which the pipes are made. The actual roughness to be taken will depend upon the quality of workmanship, the accuracy with which joints are centred, and the build-up of slime or grit which the designer may think reasonable before the sewer requires maintenance' [1].

Hydraulics Research Papers Nos. 2 and 4 [4, 5], originally published in 1963, put forward suitable values for k_s for use with the Colebrook–White formula as 0·6 mm, 1·5 mm and 3·0 mm, according to whether the sewer condition was 'good', 'normal' or 'poor' respectively. A later paper by Ackers et al. [18] suggested that those values of k_s in the Colebrook–White formula made suitable allowance for the build-up of slime in drains and sewers constructed of *any* material. The k_s value of 1·5 mm over the lower 25% of the pipe surface is more or less equivalent to a coefficient of $n = 0·012$ in the Manning formula. The Working Party Report [17] has recommended that the roughness figures quoted in the Hydraulics Research Papers [4, 5] should continue to be used pending further research.

The values given for velocities and discharges in the Crimp and Bruges tables (being based on a Manning coefficient of $n = 0·012$) therefore satisfactorily take into account the effects of the build-up of slime and grit in sewers and drains after a short period of use. It follows that these tables [39] can therefore be used for pipes of any materials. For larger pipelines (over about 900 mm diameter) and for culverts, the values of n to be used with Manning's formula can be taken from Table 5.2.

Kutter's formula is commonly used in the U.S.A. for the solution of problems involving flow in sewers. The formula is rather unwieldy but its use has been simplified by the preparation of graphs in many textbooks. The n coefficient in the Manning formula is sub-

TABLE 5.2

FRICTION COEFFICIENTS FOR USE WITH MANNING'S FORMULA

Material	Coefficient n
Cast-iron pipes	0·013 to 0·017
Concrete	0·015
Well-planed timber	0·008
Good brickwork or stonework	0·015
Old brickwork or stonework	0·020
Corrugated metal culverts	0·021
Trimmed earth, canal banks, etc.	0·025 to 0·030

stantially the same as the *n* used in the Kutter formula (except for corrugated pipes), and, in view of its simplicity, the Manning formula is more widely used in the U.K.

VELOCITY OF FLOW

The velocity of flow in a sewer or open channel must be sufficient to prevent the deposition of solids. If the velocity is insufficient, solids will settle to the invert of the sewer and will remain there. They will only be returned to suspension when the velocity of flow increases sufficiently to cause turbulent flow conditions.

As stated above, it is usual to design surface water sewers so that the velocity of flow when full is not less than 0·75 to 1·00 m/s. Earlier designs were often based on a *maximum* velocity of from about 1·8 to 3·0 m/s, as it was felt that excessive velocities would cause erosion of the pipe materials by the grit-laden sewage. Experience in recent years has shown, however, that the scouring effect of sewage is less serious than was once thought; as the construction of sewers at steep gradients can be considerably cheaper than the construction of backdrop manholes, an upper limit on velocity to avoid scour is now rarely adopted.

The above figures for minimum self-cleansing velocities are applicable to small and medium sized sewers (up to about 900 mm diameter). In larger sewers the effect of turbulence is less marked and there is a greater tendency for grit to be deposited; in these circumstances the gradient should be chosen to provide higher velocities if possible (i.e. minimum velocities greater than 1·0 m/s).

GRADIENTS

The diameter and gradient of a sewer are therefore chosen to give a satisfactory full velocity. In some cases the gradient will be set by site conditions, so that a suitable diameter of pipeline will be chosen to suit the gradient. In other cases, where the gradient is not critical, it may be possible to choose a small diameter of pipeline (for economy) and then to adopt a gradient to suit. Many authorities in the U.K. now consider that the minimum diameter for a surface water pipeline should be 225 mm, with 150 mm diameter branches for connections from individual properties, gullies, etc.

When the Crimp and Bruges formula is employed, it is a comparatively simple procedure to choose a suitable gradient from sets of tables [39] or from the chart in Fig. 5.1.

If a pipeline is not surcharged, the hydraulic gradient will normally

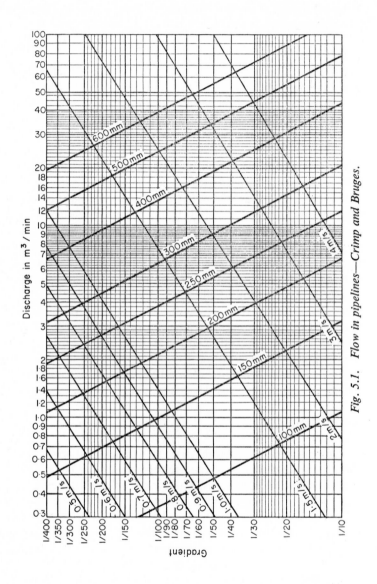

Fig. 5.1.   Flow in pipelines—Crimp and Bruges.

be more or less parallel with the line of the pipes. It is general practice, therefore, to design the gradient of the *invert* of a drain or sewer and to assume that it will not flow surcharged. On the same basis, the soffit of a pipeline will represent the hydraulic gradient when the pipe is running full, but not surcharged. For this reason, the lines of soffits of sewers should be continuous, and any steps due to increases in diameter of the pipes should be formed in the invert. In practice this means that the invert of a 225 mm pipeline discharging out of a manhole should be set 75 mm lower than the invert of any 150 mm pipeline discharging into the manhole.

In the past, considerable use was made of McGuire's Rule for drainage in the immediate vicinity of buildings. This laid down minimum gradients of 1 in 60 for 150 mm pipes and 1 in 90 for 225 mm pipes. The rigid adoption of this rule (devised over 100 years ago), without due consideration of other factors, has often resulted in bad drainage design where upper sections of drains have been extremely shallow, where unduly steep gradients have resulted in uneconomic depths of sewers or where the pipes have been of a larger diameter than necessary. In view of the present-day knowledge of self-cleansing velocities (and gradients) and the availability of tables and graphs, there is now no excuse for the perpetuation of this extravagant rule-of-thumb method.

Equally extravagant is the tendency of some designers to specify a sewer of *larger* diameter than required so that the gradient appears (on paper) to give a self-cleansing velocity. The velocity of flow is, of course, dependent on the quantity, or rate of flow, and not on the pipe diameter. For the same gradient, at small rates of flow, a small diameter pipeline will be more self-cleansing than a larger pipeline.

FORMULAE FOR THE FLOW IN PIPES
The Manning formula is

$$v = \frac{0 \cdot 01}{n} r^{\frac{2}{3}} s^{\frac{1}{2}} \qquad \textbf{Formula 5.1}$$

where

v is the velocity of flow, in m/s,
n is a coefficient of friction,
r is the hydraulic mean depth, in mm,
s is the hydraulic gradient.

For pipes running full or half-full, this can be expressed as

$$v = \frac{0.004}{n} d^{\frac{2}{3}} I^{-\frac{1}{2}}$$ **Formula 5.2**

where

    d is the diameter, in mm,
    I is the inclination.

When n is taken as 0·012, the Crimp and Bruges versions of the Manning formulae are

$$v = 0.33 \, d^{\frac{2}{3}} I^{-\frac{1}{2}}$$ **Formula 5.3**

$$Q = 26 \times 10^{-8} \, d^{\frac{8}{3}} I^{-\frac{1}{2}}$$ **Formula 5.4**

where

    Q is the discharge, in cumec.

The Crimp and Bruges formulae are used extensively for sewer design, and the coefficient of 0·012 has been found to be generally satisfactory for 'used' sewers. For larger pipelines and culverts, the coefficient n may be taken from Table 5.2. The tables and diagrams by Crimp and Bruges, which were published originally in 1897, are based on these formulae and are widely used. They have been revised, and a metric version was published in 1969 [39]. The Crimp and Bruges formulae for velocity and discharge are set out graphically in Fig. 5.1.

FORMULAE FOR THE FLOW IN CHANNELS

Formulae used for the flow in pipelines are equally applicable to open channels.

    The Chezy formula for the flow of water is

$$v = C \sqrt{mI}$$ **Formula 5.5**

where

    v is the velocity of flow, in m/s,
    C is a constant,
    m is the hydraulic mean depth, in m,
    I is the inclination (i.e. friction head divided by length of pipeline or channel).

Values of C for various types of channel (for use with formulae in S.I. units) are given by the Bazin formula:

$$C = \frac{87}{1 + \dfrac{k}{\sqrt{m}}}$$ **Formula 5.6**

where

k has the value given in Table 5.3.

TABLE 5.3
FLOW IN CHANNELS
(Values of k for use with the Bazin and Chezy formulae—S.I. units)

Smooth cement finish; planed wood	0·06
Clean brickwork, stone, planks, etc.	0·16
Dirty brickwork, stone, planks, etc.	0·28
Rubble masonry	0·46
Earth channels in excellent condition	0·85
Earth channels—generally	1·30
Earth channels—very rough or with weeds	1·75

Having calculated the velocity of flow in a channel, the capacity can be calculated from

$$Q = vA$$ **Formula 5.7**

where

Q is the discharge, in cumec,
A is the cross-sectional area of flow, in m^2.

The original Chezy formula was used by Manning (with a suitable coefficient) to arrive at his formula, which in turn formed the basis of the Crimp and Bruges formula.

For rectangular channels constructed of concrete with a smooth finish, *when the depth of flow and the width of the channel are equal*, the flow can be calculated from the following two formulae. These are based on the Crimp and Bruges formulae, as modified by Barlow [34]:

$$v = 40\, w^{\frac{2}{3}}\, s^{\frac{1}{2}}$$ **Formula 5.8**

where

> v is the velocity of flow, in m/s,
> w is the width of the channel, in m,
> s is the ratio of the fall to length.

$$Q = 40 \, w^{\frac{3}{2}} s^{\frac{1}{2}}$$ **Formula 5.9**

where

> Q is the flow, in cumec.

Tables for calculating velocities and discharges for other depths of flow are given in 'Wastewater Treatment' [36]. The maximum discharge of a channel occurs as follows:

(i) *Rectangular*—when the depth of water equals half the breadth of the channel, i.e. $d = b/2$;

(ii) *Trapezoidal*—when the two sides and base are tangential to a semicircle drawn with its centre on the water line, or when the hydraulic mean depth equals half the depth of flow.

Maximum permissible velocities for open unlined ditches and canals were put forward by Fortier and Scobey in a paper in 1926 [24]. The metric equivalents of some of the figures given in that paper are set out in Table 5.4.

PROPORTIONATE VELOCITY AND DISCHARGE

Calculations for flows and velocities at depths other than full flow conditions are more relevant to the design of foul sewers than to surface water sewerage. There are, however, occasions when this information is required in surface water sewer design, e.g. when a sewerage system is to be constructed ahead of development and will therefore operate at less than design flow over a period of time. The relationships between velocity, discharge and depth of flow are given in Table 5.1.

SURCHARGE IN SEWERS

Sewers are designed for specific conditions of flow. A surface water sewer is designed to take the run-off from a specific storm condition (once per year, once every 2 years, etc.). Any flow in excess of the designed flow will result in a surcharge of the system, with the hydraulic gradient rising above the actual gradient of the pipeline. i.e. rising above the soffits of the pipes.

TABLE 5.4

PERMISSIBLE VELOCITIES IN UNLINED DITCHES AND CANALS

Material excavated for canal	Velocity after ageing (m/s)		
	Canals carrying clear water	Water transporting colloidal silts	Water transporting non-colloidal silts, sands, gravels or rock fragments
Fine sand	0·45	0·75	0·45
Sandy loam	0·55	0·75	0·60
Silty loam	0·60	0·90	0·60
Alluvial silts when non-colloidal	0·60	1·10	0·60
Ordinary firm loam	0·75	1·10	0·70
Fine gravel	0·75	1·50	1·15
Stiff clay	1·15	1·50	0·90
Alluvial silts when colloidal	1·15	1·50	0·90
Coarse gravel	1·20	1·80	2·00
Pebbles and shingles	1·50	1·70	2·00
Shales	1·80	1·80	1·50

Surcharge in a sewer produces a pressure, or 'head', in the system, and results to some extent in an increase in the rate of discharge and also provides some additional storage within the system. The effect of the latter is considered in more detail in Chapter 7. As sewers are generally laid a metre or more below the surface, it will be apparent that most systems can accept some surcharge before this becomes apparent at the surface. Eventually, if the amount of surcharge increases, the effect will show in the lifting of manhole covers under the pressure of the contained water, and in flooding at road gullies. While a system should rarely be *designed* to surcharge, the effect of surcharge on the capacity of a system should always be borne in mind when considering the possibilities of flooding.

CHAPTER 6

Storm Overflows

When there are separate sewers for foul and surface water, the foul sewage generally flows to a treatment works and the surface water is discharged to suitable streams or rivers without any treatment. Modern practice is now to provide separate sewers for foul sewage and for surface water, but many older sewerage systems are either combined or 'partially separate' (i.e. carrying a proportion of the surface water run-off along with the foul sewage). These older sewers have frequently developed from the culverting of ditches and streams, and are often basically surface water sewers into which foul sewage has been admitted as the science of waterborne sewerage has developed.

If only one combined sewer has been provided for an area of development, that sewer will discharge to the treatment works, and as flows during times of storm are often up to 20, or may even be over 100 times, the dry weather flow, one or more overflows may be necessary to regulate the flow to the works in times of storm. These 'storm sewage overflows' then pass diluted sewage to a stream or river, either directly, or by way of storm tanks or through special storm relief sewers. No overflow should be constructed where the discharge from it would be into a watercourse used as a source of potable water, unless the conditions are such that the river is self-purifying before the water intake is reached. At least 8 km of river should normally be allowed for this. Special considerations may also apply where an overflow discharges into a watercourse used for industrial water supply, fishing, boating, bathing or other public recreation.

Storm overflows were often designed in the past to divert all flows over 'six times dry weather flow', irrespective of either the dilution afforded by the receiving stream or the capacity of the sewers downstream. More recently there has been a tendency to reduce the number of overflows to a minimum, and to set the levels of overflow in terms of actual rates of flow rather than on theoretical dry weather flow rates. It is not normal to provide a storm overflow on a sewer of less than 450 mm diameter unless this is at a pumping station or at a treatment works.

Overflows should generally be sited as near as possible to the point of discharge to a stream unless a storm relief sewer is to be provided. The outlet at the watercourse should discharge to a concrete apron near the bed of the stream, complete with suitable wing walls. A flap valve or a bar screen should be provided to prevent the ingress of rats or children; many engineers prefer a flap valve as this offers less chance of blockage by debris.

Overflows should be designed so that they only discharge when either the downstream sewer or the treatment works is incapable of accepting the total flow, as there is no virtue in overflowing storm sewage merely in order to conform with a formula. The storm overflows should be constructed at their final setting from the start. The only exceptions which should occur are where the practice would entail the premature installation of pumping plant or where an overflow on the sewerage system would be preferable to one at the treatment works; this latter case is very unlikely [9].

The sewage in combined sewers is always more polluted during the early stages of a storm; it will not only contain washings from roads and other surfaces but also pockets of accumulated sludge and silt from the sewers which have been dislodged by the increased velocity of flow due to the storm. Any overflow should be designed to reduce the pollution from these 'first flushes', and, as far as possible, the first flush of storm water should be passed forward for treatment so that the liquid which overflows will contain the minimum of polluting matter. In a report published in the U.K. in 1970 [7] it was recommended that storm sewage should be screened before discharge to a stream; at a larger installation these screens can be fitted with automatic raking arrangements.

GENERAL PRINCIPLES OF DESIGN
Ideally [10], a storm overflow should achieve the following:

   (a) it should not come into operation until the prescribed flow is being passed to treatment;

(b) the flow to treatment should not increase significantly as the amount of overflowed storm sewage increases;

(c) the maximum amount of polluting material should be passed to treatment;

(d) the design should avoid any complication likely to lead to unreliable performance;

(e) the chamber should be so designed as to minimise turbulence and risk of blockage—it should be self-cleansing and require the minimum of attendance and maintenance.

In addition, the capital cost of construction should be kept to a minimum.

The strength of the storm sewage discharged at an overflow will vary with the strength of the dry weather sewage and with the time of day; it will also be affected by the intensity of the storm, the length of time that has elapsed since the start of the storm and the time interval since any previous storm. It will, of course, be appreciated that the design of a storm overflow can make little or no provision for these variables.

To take account of the population contributing to the foul sewage flow, plus an allowance for normal infiltration and industrial flows, it has been suggested recently [10] that storm overflows should be set so that they only discharge when the flow in the incoming sewer is greater than

$$\text{d.w.f.} + 1360P + 2E \text{ litres/day} \qquad \textbf{Formula 6.1}$$

where

d.w.f. is expressed in litres/day,

P is the contributing population,

E is the volume of industrial effluent (in litres) discharged in 24 h.

In the above, the d.w.f. figure is made up of the addition of three items:

PG, which is the product of population and the average domestic water consumption per day;

I, which is the daily rate of infiltration; and

E, which is the daily industrial flow, as before.

It is recommended that where an overflow is to be provided on a sewer which serves an area where a part of the sewers is on a 'separate' system, some suitable allowance should be made for that area in the calculations. Engineers generally agree that the setting of the over-flow should be raised to accommodate an additional amount equal to three times the d.w.f. (i.e. 3*PG*) from the separate area, provided that Formula 6.1 is based on the population of the remainder of the area. While this is not strictly correct mathematically, it makes allowance for the normal three d.w.f. capacity of any treatment works down-stream of the overflow.

Where flooding occurs due to the overloading of a combined sewer, and an overflow is to be provided to discharge to an intercepting sewer, that sewer will normally be reconnected to the original sewer at some point downstream where the capacity is adequate. In that case the setting of any overflow weir or siphon will be based on the capacity of the downstream sewer and not on any considerations of contributory population or dry weather flow.

TYPES OF OVERFLOW

Traditionally, the overflow of storm sewage has been arranged either by means of side weirs or leap weirs; the latter involved quite com-plicated hydraulic problems and are now rarely used. More recent developments to regulate the rate of overflow have included the use of stilling ponds and vortex overflows, associated with orifice control of the flow in the d.w.f. channel, and of siphons.

Side weirs

The simplest form of side weir can be formed on a sewer by cutting away the sides of a length of the pipe where it passes through a chamber. Where the weirs are below the horizontal diameter of the pipeline, these are usually referred to as *low* side weirs. There is no loss of hydraulic head with this low type of side weir, but as the flow in the sewer varies there is little or no control of the rate of overflow or of the amount retained in the sewer; this type of overflow is therefore now rarely constructed, although a number of older overflows are of this type.

A chamber incorporating either one or two *high* side weirs above the horizontal diameter of the pipeline is probably the most common form of overflow constructed on medium and smaller sized sewerage schemes. The weir should be as long as economically possible, and the overflow level should preferably be near the soffit of the pipeline; this type of weir retains the advantage of operating with no significant loss of head.

The weir level should be adjustable to allow for increases in the contributing population to the sewer unless this adjustment can be incorporated into the d.w.f. outlet of the chamber. Dip plates (scum boards) must be provided between the line of flow and the weirs to retain floating solids in the main line of flow. When tested in 1970 [10], the high side weir type of overflow, fitted with scum boards, had the best performance of all overflows in retaining gross solids and faeces.

It is a well-established fact that the capacity of a circular pipeline, under gravity flow conditions, is the same when running at 0·82 of its depth as when running full; in between, the capacity increases to about 107·6% of its full value. It will therefore be apparent that if overflow weirs are set at 82% of the downstream depth, and if the maximum depth at the weir is never more than 100% of that depth, the sewer below the overflow will never be surcharged, and the maximum flow in that sewer can therefore be calculated quite accurately. The length of the weirs at the overflow should therefore be chosen to discharge the excess flows within these variations of head. The length of weir must be relatively long if close regulation is to be obtained; this is the main reason for the adoption of double-sided weirs.

The Coleman–Smith formula for calculating the length of side weir required is as follows:

$$L = 14 \cdot 9 \; W v H_1^{0 \cdot 13} \left(\frac{1}{\sqrt{H_2}} - \frac{1}{\sqrt{H_1}} \right) \qquad \textbf{Formula 6.2}$$

where

L is the length of weir required, in mm,
W is the mean width of the main channel, or the distance between dip plates, in mm,
v is the mean velocity in the channel, in m/s,
H_1 is the incoming head above the weir, in mm,
H_2 is the outgoing head above the weir, in mm.

Earlier formulae took into account one value only for the head on the weir (H). The purpose of a well-designed weir is, however, to reduce the outgoing head (H_2) to a minimum and this is often taken as about 20 mm.

When using the Coleman–Smith formula it should be remembered that it has certain limitations. While the velocity along the channel is known to vary, no account is taken of this in the formula. The investigations of Coleman and Smith were based on a channel of constant width with unrestricted outlet, and the application of the formula to other conditions may introduce errors. It is usual to add 10% to the calculated length of the weir if this is to be broad-crested, and a further 10% if scum boards are to be used.

Two formulae prepared by Engels (and adjusted for the use of metric units) are:

(i) *For side weirs on a straight channel*

$$Q = 1 \cdot 833 \, L^{0 \cdot 83} \, h^{1 \cdot 67} \qquad \textbf{Formula 6.3}$$

where

Q is the discharge, in cumec,
L is the length of weir crest, in m,
h is the head on the weir, in m.

(ii) *For a contracted channel*

$$Q = 1 \cdot 833 \, L^{0 \cdot 9} \, h^{1 \cdot 6} \qquad \textbf{Formula 6.4}$$

where

Q is the discharge, in cumec,
L is the length of weir crest, in m,
h is the head on the weir, in m.

It will be noted that the ratio of channel width to contracted width does not affect Formula 6.4.

Baffled side weirs
For a normal side weir overflow the length of the weirs must be relatively long. To overcome this experiments have been carried out with a baffle placed across the main channel; the open area below the baffle can be treated as an orifice when calculating the probable flow downstream of the overflow. The induced pressure head will result in an increase in velocity in the outlet pipe and eventually a rise in level downstream as conditions return to normal gravity flow.

A 'controlled outlet' is not practicable when *low* side weirs are used as this can result in the formation of a standing wave; the depth of flow at the control is not then directly predictable.

Some storm overflows have been constructed with the outlet restricted to a semi-circular cross-section. The capacity of the half-round channel can be calculated by normal formulae until just before the flow reaches the horizontal crown; when this is reached the discharge and velocity of flow will be substantially reduced. As the new hydraulic mean depth is known, the surcharge required to restore the flow to its earlier rate can be calculated; this will determine the level for the overflow weirs.

Transverse weirs

A transverse weir can be constructed across the line of flow in a chamber so that normal flows continue under the weir, with the diverted flow over the weir being directed to one side. This allows a more accurate calculation of the discharge over the weir as there is no variation in head *along* the weir (as with a side weir) but the increase in depth upstream of the weir will result in a surcharge of the downstream sewer in times of maximum run-off. As far as the author is aware, this type of overflow has not been used in the U.K. although it is referred to in American literature.

The general formula for a broad-crested weir is

$$Q = 1 \cdot 71 \, BH^{1 \cdot 5} \qquad \text{Formula 6.5}$$

where

Q is the flow, in cumec,
B is the length of weir, in m,
H is the head on the weir, in m.

Leap weirs

Leap weirs were used to some extent in the past, but are now rarely encountered. Normal low flows drop from the combined sewer through an opening in the invert into a sewer which discharges to the treatment works; at times of high rates of flow (and high velocities) a proportion of the flow then 'leaps' over the aperture leading to the sewer and falls onto an apron which discharges to the overflow. Adjustment of the overflow rate is possible by means of a plate at the weir, but this is very much a matter of trial and error.

Stilling ponds

It is generally accepted that the effect of the highly polluting first flushes are minimised by the use of either high side weirs or by the relatively modern development of the stilling pond. Some element of storage is then provided before the overflow comes into operation. The overflow outlet from a stilling pond can be by side or end weirs or by a siphon. The primary function of a stilling pond is the dissipation of the energy of the incoming sewage and the creation of flow patterns which will lead to the separation of the solid particles so that these are not carried over with the storm flows.

The stilling pond is in effect a length of V-shaped sewer, the apex of the V being below the invert of the main sewer and sloping towards the outlet downstream. The outlet consists of a length of small diameter pipe by means of which the flow is controlled. In dry weather the flow along the apex is self-cleansing, but during storms the V fills up and a stilling pond is formed in which most of the energy of the approaching stream is absorbed, so that by the time the stream approaches the overflow weir placed across the downstream end of the stilling pond it is more or less quiescent. Scum and other floating material, therefore, rise to the surface and are trapped, and in due course when the storm subsides this material passes on to the treatment works.

When a weir is used to control the overflow, the stilling pond may take the form of an enlarged manhole chamber. The weir is then constructed across the main channel and over the smaller diameter outlet pipe so that the crest of the weir is above the soffit of the inlet sewer; this allows a pond to be formed in times of storm, before the storm overflow outlet comes into operation. In some circumstances more than one weir may be necessary to give the required length.

The results of experiments carried out in recent years in the U.K. suggest that where a siphon is used this should preferably be located near the downstream end of the stilling pond chamber, where the flow is less liable to vortex formation and where the solids separation is more efficient.

Whether a weir or a siphon is used, a baffle should be incorporated so that floating solids are retained in the pond; these will then discharge through the smaller diameter d.w.f. outlet after the storm has subsided. Turbulence in the pond will be reduced if the gradient of the last section of sewer upstream is kept as low as possible; this must, of course, not be reduced below that necessary to give a self-cleansing velocity at dry weather flow.

Vortex overflows
A further recent development in storm water overflow design is the vortex. The author is only aware of one sewerage system in the U.K. which actually incorporates vortex overflows. From research carried

out it is apparent that the use of this type of overflow could result in a reduction of the amount of pollution carried to the overflow outlet.

A vortex overflow consists of a circular chamber and a peripheral channel. The incoming sewage enters tangentially and the dry weather flow element leaves through a smaller pipe set tangentially to the central circular weir. This weir has its cill below the main sewer soffit level and forms the basis of the overflow outlet.

The passage of the sewage flow around the peripheral channel results in the heavier solids moving along the floor towards the centre (and therefore being discharged via the d.w.f. outlet) while the lighter floating solids move towards the outer wall. A dip plate prevents floating solids being carried over the weir to the outlet.

Siphons

Any difficulty due to the size of the overflow needed to accommodate the length of weir normally required can be overcome by using a siphon overflow. A compact structure can then be formed using one or more siphons with their outlets well above sewer invert level and below top water level so that, as far as possible, floating solids are not discharged through the siphon; care must, however, be taken that accumulations of scum are not sucked through the siphon just before it deprimes. The siphon operates with a much smaller variation in water level than is required for a weir overflow.

The design of siphon overflows is more complex than for weirs, but recent research has resulted in increased knowledge of their operation. As the use of a siphon is particularly appropriate when associated with a stilling pond, there is a tendency to use siphon overflows, particularly on larger schemes.

The approximate cross-section of the throat of a siphon can be calculated from the following formula:

$$Q = mA \sqrt{2gH} \qquad \qquad \textbf{Formula 6.6}$$

where

Q is the discharge, in cumec,
A is the area of the throat, in m^2,
m is the coefficient of discharge, normally between 0·6 and 0·8,
g is 9·806 m/s^2,
H is the head, in m.

ORIFICES AND THROTTLE OUTLETS

The outlet for the d.w.f. element from a stilling pond, vortex or high side weir overflow is probably best controlled by some form of orifice or throttle pipe to restrict the flow to the determined maximum. Where further development is expected, this orifice can then be made adjustable.

The available head will determine the cross-section of this outlet, and the maximum discharge should preferably not exceed the design discharge by too great a margin. The shape of the orifice should, of course, also be such that solid matter will pass without blocking. The U.K. Technical Committee on Storm Overflows and the Disposal of Storm Sewage [10] recommended that any orifice outlet should have a minimum dimension in any direction of 150 mm, and a minimum cross-sectional area of 23×10^3 mm^2.

The discharge from an orifice can be calculated from the following formula:

$$Q = mA \sqrt{2gH} \qquad \text{Formula 6.7}$$

where

Q is the discharge, in cumec,
A is the area of the orifice, in m^2,
H is the head, in m,
g is 9·806 m/s^2,
m can be taken as follows:

sharp-edged plate	0·62,
penstock	0·62,
opening in wall of chamber	0·86,
short length of pipe	0·81.

STORM TANKS

Storm sewage tanks are generally constructed alongside any sewage treatment plant so that their contents can be returned to the works for treatment at the end of a storm. In a few recent installations, tanks have been constructed at storm overflows on trunk sewers *before* the main flow reaches the treatment plant. In the past, these tanks have generally been designed to give a minimum settlement time of 2 h at maximum flow (up to six times d.w.f.). It has been recommended [10] that, in future, tank capacity should be provided, for normal conditions, on a basis of about 65 to 70 litres per head of contributory population.

STORM RELIEF SEWERS

A storm relief sewer may be a sewer carrying the discharge from a storm overflow to a watercourse or it may be a new sewer constructed as an intercepting sewer parallel to an original sewer, to provide additional capacity in times of storm. In the latter case, if the sewer does not reconnect to the original sewer at some point downstream, it will discharge either to a treatment works or to specially constructed storm tanks.

Where storm overflows have to be provided on combined or partially separate sewers at points remote from a watercourse, a storm relief sewer can be constructed to collect one or more of these discharges; the outlet to the watercourse can then be sited to give the least nuisance from any possible pollution and may also incorporate storm tanks to further reduce the pollution load in the flow to the river. A storm relief sewer will not, of course, carry any flow in times of dry weather.

CHAPTER 7

Storage

Much urban development throughout the world has tended to be along river valleys (for ease of transport, water supply and irrigation), and most of the problems relating to flooding concern its alleviation in the built-up areas alongside rivers. Earlier works usually included *levees*, or flood banks; more recently the practice has been to control flooding by a combination of storage reservoirs and channel improvements.

As towns and cities develop, the situation can arise more and more frequently where the capacity of a river or of a sewerage system is insufficient to carry the increased discharge which would result from an area of development; in some circumstances restrictions such as culverts under railway embankments may already cause local flooding. The run-off factor from agricultural land is usually about 10 to 15%, whereas if this is developed for housing the considerable increase in paved and roofed areas may increase this run-off to 35% or in some cases as much as 45%; with industrial development this factor can be as high as 80 to 100%. Unless some form of balancing is introduced, the result of more than doubling the rate of storm run-off, together with a reduction in the time lag between rainfall and run-off, may entail the provision of a new system of main sewers downstream or improvements in the capacity of watercourses. An increase in the maximum velocity in an open watercourse, and the disturbance of the natural *regime* of the stream, may result in erosion of banks and other damage.

In some circumstances, where urban development is at the lower part of a larger catchment, it may be possible to provide balancing arrangements on the existing watercourses *upstream* of the development, so that the maximum run-off through the developed area will be reduced. At McGregor (Iowa, U.S.A.) about three-quarters of the developed area lies in a system of valleys with a high run-off coefficient and when the drainage system was being improved some years ago a series of retention dams was constructed in the tributary valleys; these resulted in a 63% reduction in the storm run-off at the lower section of the main valley sewer, thereby reducing the required capacity (and therefore the cost) of the proposed sewers within the developed area.

The construction of a balancing reservoir downstream of any proposed development will provide an economical method of controlling the run-off to within limits which are acceptable to the existing system, as arrangements can be made to store a part of the flow being discharged through the surface water sewers during times of peak flow, that part being returned to the downstream sewers or stream when the peak flow has subsided. Sharpe and Shackleton [32] have stated that 'the minimum quantity of water compatible with the required reduction in peak flow should be stored for the shortest period. If this condition is fulfilled, the storage capacity of the reservoir will be used to maximum advantage. In addition, the reservoir will be empty at the earliest possible moment and thus be available to store any excess flow from a subsequent storm'.

Balancing reservoirs have been in use on sewerage systems in the U.K. and overseas for 20 years or more; these reservoirs have been given many other names such as 'water meadows', 'retention basins', and 'storage tanks'. Some designers prefer to distinguish between water meadows (or flood meadows) as those which normally contain no water, and balancing reservoirs (or storage tanks) as those which contain water at all times; in the latter case the balancing is then a matter of variation in water level in the reservoir. In general the author has preferred to use the water meadow type of balancing as it is relatively cheap to construct and maintain, the danger of drowning is considerably reduced and the area can be used for recreation during dry periods.

Earlier development of balancing reservoirs often involved long and tedious design calculations which were more applicable to large river control schemes than to the development of comparatively small areas. Increased knowledge of the principles involved has led to simplification of the calculations. Balancing has proved particularly popular in the U.K. at new towns and at towns scheduled for extensive expansion, as the problems of flood capacity downstream are then particularly acute. In the author's opinion, much more use could be made of water meadows and balancing reservoirs for comparatively small areas of development, or for the phased development of larger areas, when the peak storm run-off would otherwise be unacceptable without the construction of expensive off-site sewers.

A considerable reduction in surface water peak flows can be effected by the incorporation of one or more reservoirs, and the greatest benefit will be obtained by providing the storage facilities as near as possible to the lower part of the drainage area as this will allow control of the maximum part of the run-off. In some situations it may be possible to make use of an existing lake on the watercourse itself by controlling the rate of overflow and by allowing for some build-up of the normal water level in times of storm. More often a reservoir or water meadow will be formed either by excavation or by banking around an existing area of flat, low-lying land.

ECONOMIC CONSIDERATIONS

When any form of balancing is proposed, it will usually be based on savings in costs (capital and annual) as against the heavy costs of constructing large sewers or of improving watercourses so that they are large enough to carry comparatively heavy and infrequent rainfall. Sometimes the saving will be so obvious that further investigation of costs of possible alternative sewerage or river works may not be warranted. Otherwise the use of a balancing reservoir should be justified, taking into account the cost of downstream sewers (with and without balancing), or the benefit to the community in terms of reduced frequency of flooding of property.

The cost of a reservoir will, of course, depend on site conditions and on the amount of excavation and banking involved; on a new development site it may be possible to utilise low-lying land or other areas unsuitable for building development. A completely underground tank may not necessarily be uneconomical in certain circumstances. It will therefore be apparent that any comparison of probable costs can only take place after an appraisal of the various possible schemes, e.g. one large reservoir or a number of smaller units; reservoirs as compared with water meadows; and the form of control proposed.

DESIGN CRITERIA

The criteria considered for balancing flows in either a surface water sewer or in a small open watercourse acting as a surface water sewer will usually fall into the following categories:

(i) *Rate of outflow.* This is usually the most important consideration, based on the capacity of the existing sewer or watercourse, and taking into account the estimated storm run-off *before* any development is commenced.

(ii) *Design storm.* This will often be decided after discussions with a water authority or local authority, bearing in mind the possible effects of any flooding on existing or proposed development. This may vary from a 2-year storm for a small balancing arrangement, to a 5-year storm for a reservoir serving a substantial area. When considering a larger watercourse draining an area of several thousand hectares, it may be wise to base any design on a 25- or 50-year storm.

(iii) *Water area and depth.* The water area of the reservoir or meadow and the rise and fall of the water surface will, to some extent, depend on (i) and (ii) above, but these will also be governed by site conditions.

THE HYDROGRAPH METHOD

The flow in a surface water sewer or a watercourse at any point resulting from a given storm can be graphically expressed as a hydrograph; a hydrograph is a curve plotted to show the rate of discharge against a time scale. When the data available is based on very limited records, the 'unit hydrograph' method is often used. This method was first described by Sherman in the U.S.A. in 1932 and has been revised and improved by many designers since then.

The unit hydrograph is the sum of a number of elementary hydrographs for the various sub-areas of a drainage area, modified to allow for the effects of routing through the sewers or channels. It has two basic assumptions:

(i) for a given drainage area the duration of run-off from rainfalls of the same duration are the same;

(ii) the volumes of surface water run-off within the same time increments are directly proportional to the total volume of run-off.

These two assumptions are of course not theoretically correct but experience has shown that they are satisfactory for catchment areas up to about 5000 km².

If the 'outflow hydrograph' is plotted to the same scale as the run-off hydrograph the volume stored for any particular storm can be measured off directly. This is illustrated in Fig. 7.1 where it is

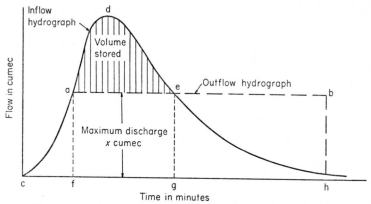

Fig. 7.1. Stormwater storage hydrograph—constant discharge.

assumed that the rate of discharge will be constant; this is of course an ideal situation which will rarely occur in practice and the discharge hydrograph will increase to some extent as the head increases to take a form similar to that shown in Fig. 7.2.

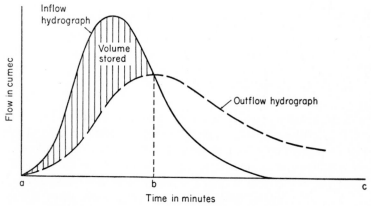

Fig. 7.2. Stormwater storage hydrograph—weir outlet.

GRAPHICAL METHODS

For smaller developments, the author has found that a 'triangle method' provides a quick and simple solution, and is of sufficient accuracy, bearing in mind the approximations that are normally built into such a scheme. The time of concentration is estimated, based on the area, shape and fall of the catchment; the relevant rainfall intensity is then calculated from the Bilham or Norris formulae for the design rainfall return period. Calculations can then

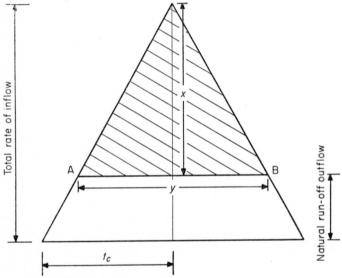

Fig. 7.3. Stormwater storage—a triangle method.

be made for (i) the 'natural' run-off before any development, and (ii) the increased run-off from the proposed development. The total of (i) and (ii) represents the maximum future run-off at the end of the time of concentration.

As all hydraulic calculations include a number of approximations, it is frequently sufficiently accurate to plot this total run-off and the time of concentration as a triangle (see Fig. 7.3). If the outflow is controlled to a constant rate to match the original natural run-off, this can be indicated by the line AB. The volume of storage required can then be calculated from the area of the triangle above AB as follows:

Total inflow (see Fig. 7.3) = 1·66 cumec
Natural outflow   = 0·68
Therefore x = $\overline{0·98}$

If t_c = 120 min

$$y = \frac{2\,(120) \times 0·98}{1·66}$$

$$= 141·68 \text{ min}$$

The volume to be stored is then:

$$0·5\,(141·68 \times 0·98 \times 60)\ \text{m}^3 = 4160\ \text{m}^3$$

In 1963, Davis [22] quoted a further simplified method of calculating the total storage required when the discharge rate is constant. The *total* run-off (Q) is plotted against duration of storm (T) as shown in Fig. 7.4, and as the discharge is constant the total discharges

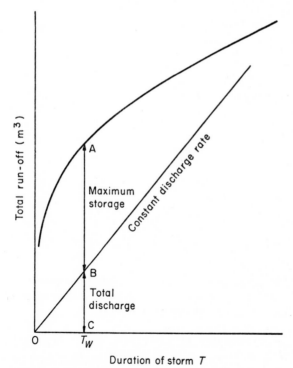

Fig. 7.4.   Stormwater storage—constant discharge rate.

at any time are represented by a straight line on the graph. The difference in the ordinates of the two curves gives the storage required for any duration, and the worst condition is that which gives maximum storage, respresented by AB, with a required total discharge BC. To allow for the time of concentration, t_c is plotted (in the same units as T, i.e. either hours or minutes) to the left of the origin and a vertical line is then erected as shown in Fig. 7.5. The tangent of the

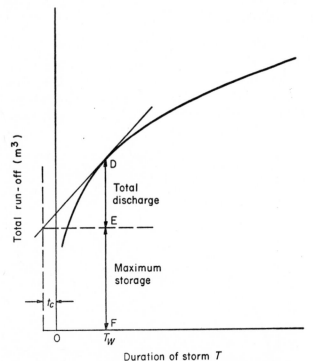

Fig. 7.5.   *Stormwater storage—constant discharge rate, with allowance for* t_c.

curve at the earlier point T_w is then produced backwards to cut the vertical line. The maximum storage required is then represented by EF and the total discharge by DE.

THE COPAS FORMULAE

In a paper in 1957 [21], Copas put forward two general formulae for the calculation of the maximum storage required for a '1-year' storm.

Escritt simplified and extended the formulae to deal with other storms. If the time of concentration can be neglected, the formula then becomes

$$C = \frac{8 \cdot 02 \, A_p^{1 \cdot 5} \, I^{0 \cdot 5}}{P^{0 \cdot 5}}$$ **Formula 7.1**

where

C is the storage capacity required, in m^3,
A_p is the impervious area, in ha,
I is the number of years between occurrences of storms (see Bilham's formula, Formula 2.4)
P is the rate of outflow from the storage area, in cumec.

OUTLETS AND SPILLWAYS

As referred to earlier, the most satisfactory form of discharge control will allow the maximum permissible discharge to operate throughout the duration of the storm, thereby reducing the required storage to a minimum. In theory this is best provided by some form of float-operated module unit, but more often than not the complicated mechanism and the need for regular maintenance tend to rule this out in favour of a less accurate device such as an orifice, a weir or a flume. While the discharge from a module can be made constant for any level of water in the reservoir, the discharge from an orifice, etc. will vary with the head. For an orifice, the discharge varies as $H^{\frac{1}{2}}$, while for a weir or flume it varies as $H^{\frac{3}{2}}$.

Discharge rates can be calculated from the following formulae:

(i) *Orifices*

The general orifice formula is set out in Formula 6.7 (p. 47).

When the orifice is a penstock or a sharp-edged plate the coefficient of discharge will be 0·62, and Formula 6.7 can then be simplified to the following:

$$Q = 2 \cdot 75 \, A \, \sqrt{H}$$ **Formula 7.2**

(ii) *Broad-crested weirs and flumes*

$$Q = 1 \cdot 71 \, BH^{1 \cdot 5}$$ **Formula 7.3**

where

Q is the flow, in cumec,
B is the width of the throat, in m (or the length of the weir, in m),
H is the head on the flume or weir, in m.

PRACTICAL CONSIDERATIONS

A reservoir which contains some water at all times can form the basis of an 'amenity area' for fishing, boating, etc., but generally the costs of construction are higher and there is a danger to children from drowning. A water meadow which remains dry for most of the year is normally available for recreation (e.g. as a children's play area) and is cheaper to construct; however, the possibility of damage to any control structures by vandals must be considered. Probably the 'reservoir' type of storage is more appropriate for larger schemes while the water meadow is, in the author's view, certainly more suitable for smaller areas of development.

Ponds should not in any event be too deep and, when an adequate area is available, it is often possible to arrange for flood meadows to flood to depths of only 300 or 400 mm; depths in excess of 1200 to 1500 mm should not normally be considered. The side slopes should be as flat as possible, not more than 1 in 2, to decrease any possibility of persons slipping when the slopes are wet. Any reservoir or meadow should be laid out so that it blends with the surrounding landscape. As any form of orifice or weir is liable to blockage from debris, provision should always be included for an emergency overflow to the downstream sewer or watercourse.

A control device installed by the author makes use of one pipeline for the overflow to and the return from the balancing reservoir. This pipe can be led out of a chamber incorporating an adjustable orifice and an emergency overflow weir. One particular installation was of a temporary nature and the chamber formed one of the manholes on the ultimate sewerage system (see Fig. 7.6).

The author considers that temporary storage reservoirs could often be constructed to serve early development, these being superseded in due course by more permanent storage facilities or by enlargements to the downstream watercourses. Temporary reservoirs may also be relevant in other circumstances, e.g. when proposals for river improvements are being considered but are not expected to be completed until some years after the development.

STORAGE WITHIN THE SEWERAGE SYSTEM

In any system of rainfall collection and disposal, storage occurs at a number of stages. These include the effects of overland flow and retention, storage in gutters, etc. and, ultimately, storage within the system of pipelines itself. It is well known that many existing surface

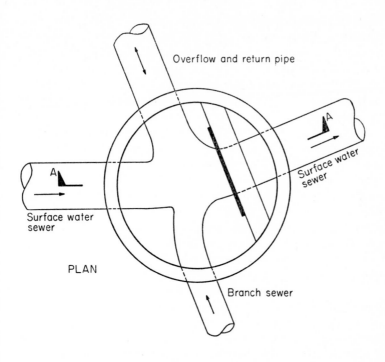

Overflow and return pipe

A

Surface water sewer

A

Surface water sewer

Surface water sewer

PLAN

Branch sewer

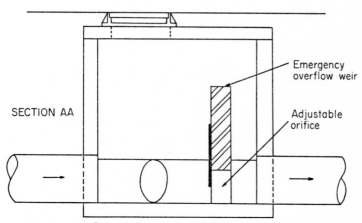

Emergency overflow weir

SECTION AA

Adjustable orifice

Balancing reservoir control chamber

Fig. 7.6.   Balancing reservoir—control chamber.

water or combined sewers which have handled intense storm flows without flooding would be classed as inadequate according to 'rational' methods of calculation. In a paper in 1955 [20], Braine showed that this ability to cope with apparent overloads could be due partly to the discharge from sewers while they are still filling, and partly to the actual storage capacity in the sewers themselves so that they continue to discharge after the storm has ceased.

The original Lloyd-Davies method and the formula associated with it (Formula 2.3) allowed for the storage which occurred in a sewerage system during a storm which would fill the sewers but which would not cause a surcharge. It has been suggested that the introduction of Formula 2.2 by the (then) Ministry of Health in 1930, while producing a rainfall curve that was reasonably statistically consistent, inadvertently introduced a new source of error in design which contributed to oversizing of sewers.

In a hypothetical case in which the rainfall is at a constant rate and continues for just as long as it takes the water to run from the top end of the longest sewer to the point of outlet, or in which the flow is to be calculated, the storage can be simply expressed. During the rainfall the rain enters the sewers at a steady rate as calculated by the formula, but the water runs out of the sewer at a rate increasing steadily from nil to the maximum rate and, therefore, at an average rate of one-half the rate of rainfall. As the storm ceases at the end of the time of concentration, the amount of rainfall stored is one-half of the total precipitated, the other half having run out of the sewer beyond the outlet. This storage in the sewer can be expressed by the formula

$$C = 30\, Qt_c \qquad\qquad \textbf{Formula 7.4}$$

where

C is the storage at the end of the time of concentration, in m^3,
Q is the rate of run-off, in cumec,
t_c is the time of concentration, in min.

CHAPTER 8

Highway Drainage

While the subject of highway drainage is specialised and is dealt with very adequately in many textbooks on road design, it is felt that a book on surface water sewerage would not be complete without a summary of this aspect of drainage design.

The *camber* on a road surface is the convex shape given to it to enable rain water to flow from the crown to the side channels; the *crossfall* is the gradient of that camber—usually taken as a minimum of 1 in 50 for asphalt, concrete and other comparatively smooth surfaces, and 1 in 35 or 1 in 40 for rougher finishes such as tarmacadam. When *superelevation* is provided at bends in roads, the convex camber is omitted and the crossfall is then usually straight from one kerb to the other.

As the area of any highway (together with any associated footways) is completely paved, the run-off should be taken as 100%, i.e. $A_p = A$. For large *unpaved* areas which are to be drained to the highway drainage system, the impermeability can be taken as between 15 and 25%. The intensity of rainfall to be allowed for in any calculation of run-off will be a matter of opinion but, in the absence of any particular local conditions, the usual basis of design in the U.K. is on a '1-year' storm; this is considered to be adequate for most highway drainage systems. The rainfall intensity can then be taken from tables or graphs based on Bilham's formula (see Fig. 2.1), or the earlier 'Ministry' formulae can be used (Formulae 2.2 and 2.3).

Where approximate figures are required, it may be convenient to work on a rainfall intensity of 40 mm/h when the time of concentration is less than about 10 min, and on 25 mm/h otherwise. Peak flows in sewers resulting from summer storms (when any unpaved surfaces are generally permeable) are hardly affected by the run-off from unpaved areas and it has therefore been recommended [13] that the *time of entry* for unpaved areas should be taken as 1 h (60 min) and the rainfall intensity taken as about 5 or 6 mm/h. This may not be the case, however, for a road in cutting, when it may be prudent to add the plan area of the cutting slopes to the area of road surface used in the run-off calculations, and to treat the whole area as impermeable.

The system of drains or sewers provided for a highway is almost

TABLE 8.1

WIDTHS OF HIGHWAYS AND FOOTWAYS

Type of road	Carriageway width (m)	Footway width (m)
Estate roads under 100 m in length	5·00	2 at 1·75
Estate roads over 100 m in length	5·50	2 at 1·75
Main access roads	6·75	2 at 1·75
Bus routes	6·75	2 at 1·75
Industrial estate roads	7·30	2 at 2·50

invariably on the 'separate' system, i.e. they are designed to carry surface water only and are not intended to carry any foul sewage. Occasionally highway engineers will insist that a highway drainage system should not be used to carry any other surface water run-off; this may be appropriate for existing highway sewers whose capacity is limited, but it is not normally an economical proposition to lay two separate systems of surface water sewers—one for the highway and one for house roofs and other areas. The usual practice in new estate development is, therefore, to provide one system of surface water sewers to carry both highway drainage and the surface water run-off from the properties.

The widths of carriageways and footways will vary with local practice. Table 8.1 may be used as a guide to the amount of surface water run-off which may be expected; this table sets out the standards set for new estate roads by a number of highway authorities in the U.K.

GRADIENTS

Opinions and specifications on maximum and minimum gradients of highways vary considerably. Without channel blocks, many highway authorities require gradients to be within the range of 1 in 15 to 1 in 120; with channel blocks, flatter gradients of up to 1 in 200 are usually acceptable. Mastic asphalt and similar materials can be laid accurately to gradients of 1 in 250.

Where a general longitudinal gradient is flatter than 1 in 200, false gradients of between 1 in 120 and 1 in 200 should be provided between gullies; this will entail the provision of false crowns between gullies to avoid having unsightly variations in the levels of the kerbs. The minimum kerb height in urban areas is usually about 100 mm, while the maximum is usually about 150 mm, as any higher kerb may cause difficulties for pedestrians, particularly those with perambulators. This difference of 50 mm will result in a maximum of 50 mm for the false crown.

The amount of crossfall and the value of the horizontal gradient are both relevant when assessing the depth of rainwater which will be flowing over a road surface during a storm; this is particularly important in the design of wide roads which will carry fast-moving traffic. Rain after falling on a road drains to the lowest level and in moving across the road surface forms a layer of water of varying thickness. This is a hazard to motorists as splash and heavy spray are thrown up by moving vehicles, while the reduced grip between tyres and wet surface increases the risk of skidding. The need to remove rainwater quickly from road surfaces has become more apparent in recent years because of the increased speed and density of modern traffic. Research carried out in the U.K. by the Road Research Laboratory has been summarised in their Report LR 236 [14]. The formula in that report (in S.I. units) is

$$d = 0{\cdot}0474 \, (L \times I)^{0{\cdot}5} X^{0{\cdot}2} \qquad \textbf{Formula 8.1}$$

where

d is the depth of water, in mm,
L is the drainage length, in m,
I is the rainfall intensity, in mm/h,
and the slope is 1 in X.

An increase in the slope of a road pavement from 1 in 60 to 1 in 30 decreases the depth of water on a road by only 11%, and it is apparent that the major benefit of a steep crossfall is the reduced volume of water which can pond in deformations in the surface [14].

GULLIES

Gullies provide the usual means of collecting rainwater which flows off the road surface; in some rural roads it may be possible to allow this water to discharge directly through 'grips' to roadside ditches. Motorway drainage is usually provided by the construction of French drains both along the complete length of the hard shoulder and in the central reservation.

A road gully comprises a steel frame and a grating seated on a gully pot which is connected by means of a 150-mm pipeline to the surface water drainage system. If the gullies are to be connected to a combined sewerage system, trapped gullies must be used to prevent any possibility of smells being transmitted from the sewers through the gullies. In the U.S.A., *in situ* catchpits of brick or concrete are generally used in lieu of gully pots.

The gully has to withstand loadings from road traffic, while the pot acts as a trap for silt and other debris. In the U.K., gully pots are available in either clayware or concrete (to B.S. 539 or 556) and gully gratings usually comply with B.S. 497; heavy duty (grade D) gratings are used at all main roads and for individual wheel loadings up to 11·4 Mg, while medium duty (Grade E) gratings are used for most other highway work. Medium duty gratings are available with straight bars (for longitudinal gradients flatter than 1 in 50) and with curved bars (for gradients of 1 in 50 and steeper). Recent research [15] has, however, indicated that curved bar gratings may not be any more efficient on steep slopes than the straight bar gratings. Care should be taken in setting the frames of gully gratings with curved bars to ensure that the vertical edges of the curved bars will face upstream when the grating is eventually fixed. Some engineers prefer to set gully gratings about 10 to 20 mm below the general road surface to assist drainage, but this is not recommended as it can produce dangerous depressions in the running surface.

A general empirical spacing provides gullies at not more than 40 to 50 m apart, or one gully for every 200 m² of impervious catchment. A formula proposed by Mollinson in a paper to the Institution of

Highway Engineers [27] recommended that gully spacing should be as follows:

$$D = \frac{280 \sqrt{S}}{W}$$

Formula 8.2

where

D is the gully spacing, in m,
S is the gradient per cent (for 1 in 25, $S = 4\%$),
W is the width of paved area, in m.

Gully spacings used by some U.K. highway authorities are set out in Table 8.2. Where a false crown of d mm has been introduced

TABLE 8.2

SPACING OF HIGHWAY GULLIES

Gradient of carriageway 1 in	Area to drain to one gully (m^2)
200	160
150	160
100	167
80	180
60	200
40	240
30	275
20	330
15	330

between gullies, the spacing between those gullies should not be more than $d/2$ m, i.e. for a false crown of 50 mm the gully spacing will be 25 m.

Care is required at road junctions to prevent ponding at radius kerbs; it is therefore preferable to site gullies just upstream of the tangent points of these kerbs to prevent surface water flowing across the junction. When a road is superelevated, a gully should be provided at the end of the superelevation just upstream of the end of the cambered section; this will avoid the possibility of water flowing across the roadway from the higher channel. The omission of gullies is false economy and additional gullies should be provided at bus stops, lay-bys, etc., and where gradients are flatter than 1 in 200 or steeper than 1 in 15.

SUBSOIL DRAINAGE

Subsoil drains may be provided in road construction and in the construction of other paved areas as a means of keeping the subgrade in a relatively dry condition. Subsoil drainage will not be effective in cohesive soils (e.g. in clay, etc.). Entry of water into the subgrade will lower the strength of the soil foundation; with non-cohesive soils it is therefore common to construct longitudinal subsoil drains at both sides of a road, with the pipes at a minimum of 1200 mm below the road formation level. These will lower the water table to a safe depth.

When a road is built into a hillside, the discharge from the adjacent uphill slopes should be intercepted into longitudinal catch-water ditches or drains to avoid any penetration of the subgrade. Similarly, care must be taken to ensure that any surface water discharged off the road is carried well clear of the subgrade.

SOAKAWAYS

Where the surface water and subsoil drains cannot be discharged to a sewer, or directly to a ditch or watercourse, it may be necessary to construct one or more soakaways. Occasionally it may be more economical to discharge to soakaways; indeed, in some situations soakaways may be *preferable* to a system of surface water sewers if a substantial increase in the run-off from a development would overload the available stream or river, or if the underground water is used for water supply purposes. Individual soakaways may be provided at each gully, etc., or a number of gullies may be connected to a sewer (normally not less than 150 mm diameter) which itself discharges to a soakaway.

A soakaway may be a pit provided with a roof slab and with open-jointed (i.e. pervious) base and sides. Frequently the pit is filled with rubble; this will obviate the need for a roof slab designed to take traffic loading, but allowance must be made for the extra capacity required to compensate for the volume of the rubble. The top of the rubble should be sealed against the ingress of soil from above. Small precast 'domestic' soakaways have capacities of up to 0·65 m³ per metre depth, while specially manufactured precast concrete segmental soakaways are obtainable, with capacities of up to 3·5 m³ per metre depth.

The usual basis of design for the capacity of a soakaway is to allow a storage volume sufficient for a minimum of 10 mm of rainfall over the area to be drained, based on 100% impermeability. As a formula (based on a rainfall of 10 mm):

$$Q = 0{\cdot}01\,A \qquad \textbf{Formula 8.3}$$

where

Q is the capacity, in m^3,
A is the area to be drained, in m^2.

It should be remembered that soakaways are only satisfactory in permeable soils or where they can be excavated down to a permeable stratum. They must also discharge above the maximum subsoil water level.

In relatively permeable soils the size of a soakaway is determined by the storage required for short, intense storms; in less permeable soils, if the design is based on a rainfall intensity for continuous rainfall (at, say, 6 mm/h), the soakaway will also be adequate for short, intense storms.

CHAPTER 9

Materials

Many schemes of surface water drainage are prepared by government or local government departments for construction by a contractor working to a formal contract. Others are prepared by developers or their advisers as part of their proposed residential or industrial development schemes. In the first category a specification will be drawn up to guide tenderers during their pricing and to control the quality of the work during construction. Standards to be adopted by residential and industrial developers will generally be controlled by specifications or conditions issued by the local authorities.

A good specification leaves a tenderer in no doubt as to the standards of materials and workmanship required by the engineer or architect. Where a contract is entered into, the specification is also a legal document and it must not therefore be open to misinterpretation; there must also be conformity between the specification, the bills of quantities and the drawings. The specification should cover all items billed and may also cover other materials or work not billed if there is reason to believe that these might be required during the contract.

STANDARD SPECIFICATIONS

Most countries have their own national standard specifications for materials used in the construction industry. Care must be exercised when quoting from any standard specification, and references must not be too general; where one standard covers a number of qualities or sizes, the quality or type required must be sufficiently described in the specification. It is general either to refer to 'the latest standard specification' in all cases (i.e. the latest at the time of tendering) or to quote the actual date of each standard. It is bad practice to mix the two methods in one specification, as this can lead to confusion during a contract. It should be borne in mind that many contractors will not have a detailed knowledge of national standards; all references to clauses of those standards must therefore be clear and unambiguous.

In British practice, when work is to be in accordance with a British Standard (B.S.) Code of Practice, all materials and components should comply with the latest edition of a British Standard Specification where applicable. The Building Regulations state that the use of any material or method which conforms with a British Standard Specification or Code of Practice 'shall be deemed to be a sufficient compliance with the requirements' of the Regulations. All pipes, manhole covers, gullies and other drain fittings must comply with the current requirements of the relevant British Standard when used on housing schemes covered by the guarantee of the U.K. National House-Builders Registration Council.

PIPELINES—GENERAL

There is a tendency in the U.K. for some local authorities to require all sewers to be adopted by them to have a minimum diameter of 225 mm; this is based on the premise that these are less likely to blockage than the more normal minimum diameter of 150 mm. This requirement was first used for foul sewerage design and is now being adopted by some authorities for *all* sewers. As the branches from private drains can be 100 mm diameter, the author is of the opinion that the argument of possible blockage of 150 mm sewers is not valid if these sewers are properly designed and constructed. The increased cost of providing 225 mm pipelines unnecessarily adds to the cost of a sewerage scheme (and therefore to the cost of any related development). As these sewers rarely, if ever, operate at their full capacity, the actual velocities of flow during normal conditions are frequently far below the theoretical velocities, with the result that blockage could be *more* likely rather than the reverse.

The more normal pipeline materials for surface water sewers are vitrified clayware and concrete. The choice of material will generally be based on costs. In the U.K., pipelines up to 300 mm diameter are generally cheaper in vitrified clayware, with concrete pipes being used for some 300 mm pipelines and for most larger diameters; in some parts of the country the division may be at 225 mm. In France and some other continental countries, concrete pipes may be used for the smaller diameters.

The use of pipes with flexible (mechanical) joints is now mandatory in the U.K. where schemes are the subject of loan sanction by central government. It is recommended that for *all* drainage pipelines the use of rigid joints with rigid pipes should be discontinued at the earliest practicable date.

The use of uPVC pipes for sewerage has increased over recent years but this has generally been restricted to the smaller diameters used in house drainage work.

VITRIFIED CLAY PIPES

Vitrified clay drain and sewer pipes have been manufactured since about 1845 and are the most widely used for the smaller diameter gravity sewers. B.S. 65 and 540 includes two types of pipe, viz. 'British Standard' quality for general use and 'British Standard Surface Water' quality for surface water sewers only. Both types of pipe are available as either 'standard strength' or 'extra strength', and both types can be supplied with flexible joints. Part 1 of the Specification relates to the pipes and fittings; Part 2 gives details of flexible mechanical joints.

Vitrified clay pipes have safe crushing test strengths as set out in Table 9.1, and their approximate weights (in terms of aggregate length of pipes per tonne) are given in Table 9.2. In addition to the diameters listed in Table 9.1, they can be supplied in diameters of 125, 175, 200 and 250 mm and also over 300 mm at special request (the

TABLE 9.1

SAFE CRUSHING TEST STRENGTH OF VITRIFIED CLAY PIPES TO B.S. 65 AND 540

	Test strength per metre of inside length					
Nominal diameter (*mm*)	Standard strength		Extra strength		Most major manufacturers	
	(*kgf/m*)	(*N/m*)	(*kgf/m*)	(*N/m*)	(*kgf/m*)	(*N/m*)
100	2 000	19 600	2 200	21 600	2 800	27 500
150	2 000	19 600	2 200	21 600	2 800	27 500
225	2 000	19 600	2 800	27 500	2 800	27 500
300	2 200	21 600	3 400	33 400	3 400	33 400

B.S. lists diameters up to 900 mm). They are available in various lengths up to about 1·5 m and are now not normally glazed.

Pipes up to 450 mm diameter can be obtained with factory-applied 'push-fit' joints. These can incorporate polyester farings and a rubber 'O' ring or they may be of polyurethane with an integral nib. As an

TABLE 9.2

WEIGHTS OF VITRIFIED CLAY PIPES

Nominal diameter (mm)	Approximate length of pipes (metres per tonne)	
	Standard strength	Extra strength
100	76	70
150	45	42
225	27	25
300	15	14
375	9	7
450	6	—

alternative, on smaller diameters, flexible joints can be obtained with plain-ended pipes using plastic push-fit sleeve couplings. Pipes can also still be obtained with plain sockets for use with sand/cement joints. For land drainage work, perforated vitrified clay pipes are available to B.S. 65 and 540; these are manufactured to 'extra strength' quality with perforations in one half of the circumference only, and in plain-ended lengths up to 1·5 m.

CONCRETE PIPES
Concrete pipes are manufactured to B.S. 556 and can be either un-reinforced or reinforced, according to the strength classification. They are available in internal diameters from 150 to 900 mm in 75 mm increments and from 900 to 1800 mm in 150 mm increments. The minimum proof and ultimate crushing test loads for the four classes of pipe are given in Table 9.3. Pipes of nominal diameters larger than 1800 mm are available for any agreed strength, and pipes of higher minimum crushing strengths are also available for any standard nominal diameter.

Pipes can be supplied in various lengths, according to manu-facturer and diameter, from approximately 1·0 to 2·5 m. A schedule

TABLE 9.3

MINIMUM CRUSHING TEST LOADS OF CONCRETE PIPES TO B.S. 556 (LOADS IN kgf/m)

Nominal diameter (mm)	Standard		Class 'L'		Class 'M'		Class 'H'	
	Proof	Ult.	Proof	Ult.	Proof	Ult.	Proof	Ult.
150	2 010	2 380	—	—	—	—	—	—
225	2 010	2 380	—	—	—	—	—	—
300	2 010	2 380	—	—	—	—	2 380	2 980
375	2 010	2 380	—	—	3 120	3 900	3 720	4 650
450	2 010	2 380	—	—	3 570	4 460	4 170	5 210
525	2 010	2 380	—	—	3 870	4 840	4 610	5 760
600	2 010	2 380	—	—	4 610	5 760	5 510	6 870
675	2 010	2 380	—	—	5 060	6 320	6 100	7 620
750	2 010	2 380	3 870	4 840	5 360	6 700	6 550	8 190
825	2 010	2 380	4 170	5 210	5 810	7 250	6 990	8 740
900	2 010	2 380	4 610	5 760	6 850	8 560	8 630	10 790
1 050	2 010	2 380	5 210	6 500	7 740	9 670	9 820	12 280
1 200	2 010	2 380	5 800	7 250	8 780	10 980	11 160	13 940
1 350	2 010	2 380	6 400	8 000	9 680	12 100	12 360	15 430
1 500	2 010	2 380	6 990	8 780	10 560	13 200	13 400	16 750
1 650	2 010	2 380	7 580	9 480	11 750	14 690	14 880	18 600
1 800	2 010	2 380	8 330	10 420	12 650	15 800	16 070	20 090

By courtesy of the Concrete Pipe Association.

of approximate weights of 'standard' quality pipes with flexible joints, quoted by two manufacturers, is given in Table 9.4.

Concrete pipes are also manufactured with ogee joints to B.S. 4101. An ordinary ogee or rebated joint is such that the joint is made within

TABLE 9.4

WEIGHTS OF STANDARD CONCRETE PIPES
(TAKEN PRO RATA TO WEIGHT OF WHOLE PIPE)

Nominal diameter (mm)	Manufacturer 'A' (kg/m)	Manufacturer 'B' (kg/m)
150	42	56
225	70	83
300	105	143
375	142	160
450	186	189
525	216	332
600	280	450
675	337	370
750	389	492
825	475	506
900	558	580
1 050	670	775
1 200	869	1 385
1 350	1 215	1 600
1 500	1 400	1 995
1 650	1 660	—
1 800	2 047	—

the wall thickness without enlargement of one end of the pipe or fitting, and is sealed by the use of cement mortar or other suitable material. Ogee joints are not watertight, and these pipes are there-fore only suitable for use as culverts and where infiltration is not a problem. They can be used for surface water sewerage work, but are not satisfactory where the flexibility of the pipeline is important. Porous concrete ogee-jointed pipes can be obtained to B.S. 1194 for under-drainage purposes.

ASBESTOS–CEMENT PIPES

Asbestos–cement pipes for sewerage and drainage are manufactured to B.S. 3656. The standard length is 4·0 m, but half-lengths (2·0 m) and quarter-lengths (1·0 m) are available to order. The Specification includes provision for pipes of 5·0 m length in the larger diameters.

The pipes are plain-ended, the joints being made with A.C. sleeves and rubber rings. The joints are flexible and allow deflection up to 8° on the smaller diameters.

uPVC PIPES

Unplasticised PVC (uPVC) pipes are manufactured to B.S. 3505 and 3506 and are available in diameters up to 300 mm. These pipes are normally supplied in lengths of 3·0, 6·0 and 9·0 m, while some manufacturers supply 12 m lengths in the smaller diameters.

Although uPVC is theoretically a rigid material, from the point of view of structural design of pipelines uPVC pipes are flexible. Joints are formed either by solvent welding or with collars incorporating rubber sealing rings. With rubber ring-type joints, the collars may be loose or the ring may fit into an integral socket formed in the pipe itself.

B.S. 4660 refers to PVC pipes for drainage purposes, manufactured in standard lengths of 1·0, 3·0 and 6·0 m. This Specification relates only to the smaller diameters; pipes to B.S. 3505 and 3506 are frequently used in the larger diameters for drainage work.

GRANULAR BEDDING

For bedding smaller diameter pipes (up to about 300 mm diameter) it is normal to use granular material which will pass a 19 mm sieve but be retained on a 4·75 mm sieve. For larger pipes some engineers prefer to use a slightly larger material with particles up to 25 mm. An excess of fines in the material may cause 'bulking' during progress of the work, with consequent variations in line, level and compaction. When material such as gravel or broken stone is used, it should be well and evenly graded and should be physically and chemically stable in the soil and ground water to which it will be subjected. Sharp-edged stones should be avoided with PVC pipes and with specially protected pipes; limestone should not be used in soils containing sulphates or acids.

Practice in bedding pipes varies considerably outside the U.K., and while some countries use granular materials or clinker, considerable use is also made of fine soil or sand for pipe bedding. To avoid settlement of trenches in highways in France they are sometimes completely backfilled with sand or granular material.

MANHOLES

Manholes were first introduced to provide facilities for the removal of accumulations of silt and grit without the necessity of breaking into the sewers themselves. Subsequently it has become the accepted practice in most countries to provide a manhole as a means of access to a sewer at each junction and at each change of direction or gradient. In view of the expense of their construction and the obstruction caused in the highway, it is now usual to limit the number of manholes to the absolute minimum. They should be spaced at up to 100 m apart whenever possible on smaller sewers, and at up to 300 or 400 m when the sewer is large enough to permit entry for inspection.

Manholes are generally constructed either in engineering brickwork or of precast concrete rings; when concrete rings are used they should preferably be surrounded with about 150 mm of concrete to ensure adequate strength to withstand traffic and to give a fully watertight structure.

On pipelines up to 300 mm diameter, to provide easy access to the sewer and to allow the handling and jointing of drain rods, rectangular brick or *in situ* concrete manholes should be at least 1350 mm long by 788 mm wide for sewer depths to about 3 m to invert (these figures have been chosen to fit standard brickwork dimensions). On larger diameter sewers, the internal widths of manholes should be sufficient to accommodate the channel plus adequate benching; for sewer diameters varying from 375 to 750 mm, the internal width should then be 1125 mm for the smaller dimensions increasing to 1575 mm for the larger sewers. For deeper sewers to about 8 m, it is normal to have an access shaft of at least 788 mm by 675 mm

TABLE 9.5

RECOMMENDED MINIMUM SIZES FOR CONCRETE RING MANHOLES

Diameter of largest sewer (mm)	Diameter of manhole rings (mm)
150 to 300	1 050
375 to 450	1 200
525 to 600	1 350
675 to 750	1 500
825 to 900	1 800

leading to a lower chamber. This lower chamber should preferably have a minimum clear height of 2·0 m above the benching, and should generally be at least 1350 mm long by 1125 mm wide.

While a concrete manhole should not be less than 1050 mm diameter, a minimum of 1200 mm is often adopted. A schedule of suggested minimum diameters for precast concrete ring manholes is set out in Table 9.5.

In the U.K., most manhole covers and frames are manufactured to B.S. 497 in Grade 'A' for traffic loading, grade 'B' for light loading and grade 'C' where access to wheeled traffic is not possible. Step irons (to B.S. 1247) are usually provided for access.

ROAD GULLIES
A gully in the U.K. will consist of a gully grating and frame, set over a gully pot which is connected by pipeline to the surface water sewer. Gully gratings and frames will be to B.S. 497, while pots can be to B.S. 539 (vitrified clay) or 556 (concrete). Gully pots can be trapped (for connection to combined sewers) or untrapped, and various patterns are available which incorporate removable mud buckets and similar devices. The normal size of pot for estate roads and similar conditions is 450 mm diameter by 900 mm deep. It is preferable to surround the pot with concrete and to provide a flexible joint on the pipeline immediately clear of the concrete surround. Practice in the U.S.A. tends more towards *in situ* concrete or brick chambers connected to the sewer and finished with an inlet grid at street level.

DRAINAGE CHANNELS
A number of patented drainage blocks and channels are available which provide a continuous line of drainage level with the surface, obviating the need for complicated crossfalls. These are particularly applicable at factory forecourts, car parks and similar surfaced areas. Some of these take the form of slotted channels, others of U-shaped channels with removable cover slabs. These blocks and channels are generally about a metre in length and can be provided in strengths to suit varying traffic loadings.

HIGHWAY CHANNEL BLOCKS
When road surfaces have gradients flatter than about 1 in 120 it is general to use precast channel blocks alongside the kerbs to provide an effective channel to carry the run-off to the gullies. Channel

blocks are manufactured to B.S. 340 and are usually 250 mm by 125 mm in section. They should be laid carefully to line and level, set on a 150 mm base of concrete at the same time as the kerbs are laid, and jointed with cement mortar.

JUNCTIONS

While connections to sewers should preferably be made at manholes, it is quite normal to allow the connections from house drains and gullies to be made at specially manufactured junctions; these should preferably be of the 45° oblique type, although curved square junctions are sometimes permitted. A connection to an existing sewer can be made by means of an oblique saddle junction, the saddle being jointed over a hole broken into the sewer. When estate sewers are constructed ahead of development it is preferable to provide junctions in the relevant positions and to seal these off where necessary with 'joinder' pipes until they are required.

CHAPTER 10

Construction

Where pipelines will traverse open country, turf should be carefully cut and lifted and retained for re-use later. Where appropriate, topsoil should be removed to a minimum depth of 300 mm and stacked separately. The extent of the removal of topsoil will depend on the site; it may be confined to the width of the trench, but sometimes it is advisable to extend this to include the strip of land to be used for stacking excavated materials. Where farm stock will use the fields traversed by the pipeline, it may be necessary to allow for temporary fencing along each side of the working width.

Where excavation is to be carried out along or across a public road, the highway authority should be consulted early to agree the line and, where necessary, to make arrangements for road closure or for any special traffic control. The construction of sewers in built-up areas presents problems quite different from those encountered in open country. Costs are much higher for such items as road-breaking and reinstatement, traffic control, avoidance or re-routing of other services, and the protection of adjacent properties.

Each pipe should be examined carefully on delivery to the site and any that have been damaged must be clearly marked and removed. Factory-applied joints on pipes should be protected according to the manufacturer's instructions. All pipes and joints should be examined again immediately before they are laid.

The centre line and top width of the trenches should be accurately set out and marked with suitable pegs, with offset pegs where necessary. A strong sight rail, painted in contrasting colours and fixed to posts, can then be erected over the centre of each manhole, with the centre line of the sewer marked on the sight rail and the rail level located at an even dimension above the proposed sewer invert. Offset sight rails will be more satisfactory when mechanical diggers are used. As a check against accidental misplacement of a sight rail, an intermediate rail should be erected on each length of sewer. Using a boning rod or traveller of relevant length, wooden pegs are then driven into the trench bottom at intervals of about 3 or 4 m at the level of the proposed pipe invert if a granular bedding is not to be used for the pipeline; when a bedding is to be used the pipes are bedded into the granular material to their correct level using the boning rods. If pegs are used they should be removed as the work proceeds. On flat gradients, or with large diameter pipes, the work should be set out and checked frequently by instrument.

When rigid joints are used they are made with a tarred hemp gasket, caulked tightly home so that it fills no more than a quarter depth of the socket, together with a stiff cement/sand mortar to fill the remainder of the space in the socket; this is often finished off with a 45° fillet. The insides of the pipes must be kept clean with a damp cloth, and a close-fitting pad should be drawn through each pipe as it is laid. When mechanical joints are used, these should be made in accordance with the manufacturer's instructions, which will vary to some extent between different pipe materials and different diameters. The annular space outside the jointing ring must not be filled with mortar as this will impede the free flexible action of the joint and may cause burst sockets; where there is a possibility of this annular space becoming filled with gravel or stones, it can be filled with puddled clay or with fine soil.

When flexible joints are used, and pipe cutting is involved, recourse to the spigot and socket or double collar joint caulked with cement mortar or similar material is generally unavoidable. The First Report of the Working Party on Sewers and Water Mains [17], published in 1975, saw no objection to this 'if the joint is efficiently devised and properly executed, and provided the principle of overall flexibility is preserved. The principle can normally be considered fulfilled if the distance between flexible joints in the length of pipeline containing the caulked joint does not exceed the regular spacing of the flexible joints in the rest of the system. This spacing will usually be equal to the standard length of mechanically jointed pipe. The cutting of pipes should be avoided as far as possible; lengths shorter than standard are usually available from manufacturers.'

When ogee-jointed pipes are used, the joint should be thoroughly cleaned before laying, and cement mortar (one part cement to two parts washed sand) should be applied to the ends for jointing so as to completely fill the joint. The pipes should then be drawn together and the outside of the joint neatly pointed. Any surplus mortar or other material should be removed from each pipe before the succeeding pipe is laid.

If mechanical joints have been used specifically to obtain flexibility in the pipeline (e.g. in areas subject to mining subsidence, or through soils which are liable to swell and shrink), any concrete protection must also be able to move in unison. Joints must therefore be incorporated in the concrete (usually at pipe joints), and before any concrete is placed the annular spaces at pipe joints should be protected to prevent the intrusion of concrete; this can be accomplished with a ring of clay or hessian.

While small diameter pipes with flexible joints can and should be jointed manually, this is not possible with larger diameters. Mechanical pulling devices will then be needed, and care should be taken to follow the manufacturer's instructions. Small diameter pipes can be trued to line and level after jointing, but larger pipes are difficult to lift and must usually be laid in their final positions as they are jointed.

Proper records must be maintained as work proceeds. Drawings must be kept up-to-date to show the work as actually executed, a day-to-day diary should be maintained, progress recorded and records maintained as the basis for payment to the contractor. In addition, records should be kept of all types of soil strata encountered and of any soil tests carried out; details kept of all pipeline tests; and 'location books' maintained to give accurate details of positions of junctions and other fittings so that these can be located from reference points on the surface at a later date.

PIPELAYING IN TRENCHES

A pipeline will normally have been designed for a specific maximum trench width (see under 'Structural Design' later in this chapter), and any proposal by a contractor to excavate beyond the design width must be approved by the engineer. The cost of any additional work which may then be required to maintain the strength of the pipeline must be borne by the contractor.

Trenches should normally be dug to a minimum width of 300 mm plus the diameter of the pipeline where this is 150 mm or more.

Additional width must be included for any timbering or sheeting. A formula used in the U.S.A. provides for a trench width of $(1 \cdot 33d + 200)$ mm, where d is the nominal diameter in millimetres. A similar formula used by some engineers in the U.K. gives a trench width of $(1 \cdot 67d + 250)$ mm, with a minimum width of 500 mm.

Excavation in rock or other firm strata may not require any support, while, on the other hand, excavation in soft fine materials may warrant the use of close sheeting. The type of support to be provided to trench sides will vary from simple poling boards, held apart by struts across the trench, to timber runners taken down as the trench is excavated. Timber sizes will vary with trench depth and the type of subsoil. Poling boards should be about 150 by 50 mm minimum, while horizontal walings should be at least 100 by 75 mm. Timber struts can be from 75 by 75 mm to 300 by 300 mm, depending on the depth and width of the excavation. Close sheeting in timber can generally be carried out in 225 by 38 mm timbers, in lengths up to about 4 m, according to trench depth.

Steel sheet piling is often driven down along the line of the trench before excavation is begun. Light steel trench sheeting is available in lengths up to about 4 m. This is more easily driven than timber and can be withdrawn and re-used. The overlapping-type sheets normally used for temporary supports to trenches have an effective thickness of 35 mm, are 330 mm wide (centre to centre), and are usually driven in groups of up to seven or eight.

When mechanical excavators are being used in any trench which requires support, whether the trench is close-sheeted or not, the cross-bracings should be placed progressively as close behind the machine as possible.

Trenches should not be opened too far in advance of pipelaying, and they should be backfilled as early as possible after the sewers have been laid and tested. When granular bedding material is used, it should normally be possible to carefully withdraw any trench sheeting as the backfilling proceeds. If support is required for adjacent structures, any sheeting may be cut off below ground level and left in position.

Excavation should commence at the lower end of a pipeline and should proceed upstream. This will allow pipe-jointing to be carried out with the sockets facing upstream, and will permit subsoil water to drain away from the working area. If necessary, temporary drains should be laid in the bottom of the excavation, leading to a sump at the lower end. Should the excavation be taken down below the required depth, the extra depth must be filled with compacted granular material (where the pipes are to be laid on a granular bed), or with concrete if a concrete bed or surround is to be used.

The quantity of water to be removed from the trenches and other excavations should be estimated in the early stages of construction so that suitable construction methods can be adopted. In many cases, the drainage of trenches to one or more sumps may be sufficient, the water level in the sumps being kept down by pumping more or less continually, using standard contractors' pumps.

When the inflow of water is too great for normal pumping methods, well-point dewatering may be satisfactory. This system consists of a series of vertical 'well-point' pipes and risers (approximately 40 or 50 mm diameter) sunk into the waterbearing stratum. These are connected through short horizontal pipes and valves to a horizontal header pipe (usually 150 mm), which is in turn connected to a vacuum pump. While one line of header pipe and risers may be sufficient, it is usually necessary to sink risers on both sides of the line of the proposed trench. The well-points are sunk into the ground by 'jetting', i.e. by forcing water through them to scour away the ground beneath the well-point.

Backfilling under roads is a structural operation. A lightly hand-tamped layer of selected material, preferably granular and not containing any large stones or other hard objects, must be laid immediately above any type of pipe to act as a shock absorber for the compacting operations in the fill above and to relieve some of the load on the pipeline. This initial lightly tamped layer should extend to 300 mm above the crown of the pipes. Subsequent filling of trenches and around manholes should be built up in layers not exceeding 150 to 230 mm (uncompacted thickness), and each layer should be thoroughly compacted before any further material is added. Compaction should aim at the same density and moisture content as those of the undisturbed soil in the trench sides, so as to avoid subsequent settlement and consequent disturbance to the road surface and to any underground works in or near the trench. While some highway authorities may ask for the filling to be of weak concrete, this is undesirable as it can be responsible for subsequent damage to both the pipeline and to the road itself. The recent Working Party report [17] referred to this practice which can lead to unnecessarily high cost of sewer construction, problems with future excavation and eventual deterioration in the riding qualities of the road pavement.

PIPELINES IN HEADINGS AND TUNNELS

Normally, where the depth to invert of a sewer is about 6 m or more below ground, it will be more economical to excavate in heading, but in most other cases it is more usual to excavate in open trench. However, when a sewer has to be laid under a main road, railway, canal or similar obstruction, it may be advantageous to avoid open cut and to construct the sewer in heading or tunnel, or to use thrust-boring. Work in heading or tunnel can be carried out throughout the 24-h day more or less irrespective of weather conditions.

A heading is possible in firm cohesive soil or in rock. C.P. 2005 recommends that the smallest size of heading for proper working should be about 1140 mm clear height by about 760 mm width at the bottom. It is often possible to employ a combination of open trench and heading when laying sewers along urban roads. For smaller diameter sewers in heading, the longitudinal timbers should be about 225 by 75 mm or 250 by 100 mm, while framing members should be 250 by 250 mm.

In all cases of work in heading, great care is needed with the back-filling and consolidation around the pipes; as it is usually necessary to do this as pipelaying proceeds, this is simplified by the use of flexibly jointed pipes set on a granular bedding. The timbering must often, of necessity, remain in position to avoid any risk of collapse, but it should be withdrawn if possible to reduce the possibility of hard spots in the bedding. Weak concrete is recommended as a back-filling material for shallow headings, although sand and other readily compactible materials have been used successfully, particularly in deeper headings where a small gap at the top of the packing will not have any serious detrimental effect. To ensure good drainage, filling of headings should proceed *downhill*.

Traditional tunnelling methods, in which the tunnel is excavated approximately to the required slope of the finished sewer, may be used for larger diameter sewers. For sewers of 1400 mm diameter and over, tunnels can be of cast-iron or concrete segmental rings, lined with concrete or brick.

Thrustboring can be used for pipes of diameters up to 2600 mm. Reinforced concrete pipes of diameters from 900 to 2600 mm are manufactured with special square-shouldered flexible joints which are suitable for jacking. There are problems in maintaining true line and level, particularly if the subsoil is variable, and the cost of thrust pits and the blocks required for the jacks can make this method expensive. In certain circumstances there can, however, be a saving in cost over more traditional methods of pipelaying, in addition to a considerable reduction in disruption to traffic flow.

BEDDING OF PIPES

Pipe bedding falls into three broad classifications:

    (i)   trench bottom;
   (ii)   granular bedding;
  (iii)   concrete protection.

Each type of bedding has been allocated a 'bedding factor' which is used in the calculation of the strength of the completed pipeline (see Fig. 10.1 and Table 10.1). In many cases, various combinations of pipe strength and bedding are capable of carrying the calculated loads. In deciding which combination to use, the designer must consider the relative overall cost of the completed sewer in each case, and the practical advantages and disadvantages of the various beddings and methods of construction in the particular conditions likely to be encountered.

TABLE 10.1

BEDDING FACTORS

Bedding factor, Fm	Bedding class	General description
1·10	Class 'D'	Hand-trimmed trench bottom
1·50	—	Thrustbores
1·90	Class 'B'	Granular bedding and backfilling to mid-diameter
2·20	Modified Class 'B'	Granular surround to a minimum of 100 mm above the pipe soffit
2·60	Class 'A'	Unreinforced concrete cradle (120°) or arch (180°)
3·40	Class 'A$_{rc}$'	Reinforced concrete cradle (120°) or arch (180°) with reinforcement of 0·4% of the concrete area
4·80	Class 'A$_{rc}$'	Reinforced concrete cradle or arch with reinforcement of 1·0% of the concrete area

The natural trench bottom (trimmed by hand after machine excavation) is suitable for the stronger rigid pipes (cast-iron) and for the smaller diameters of concrete and vitrified clay pipes. The trench bottom should be trimmed to correct level and gradient immediately before the pipes are to be laid; this can be done by 'boning in' a traveller over the sight rails. Socket holes should be formed at each socket position, leaving the maximum length of support for the pipe barrels. Each pipe is then laid individually to line and level, using the sight rails. It is generally considered that the natural trench bottom should only be used for pipe bedding where dry conditions can be achieved and where the subsoil is such that accurate hand trimming is practicable. The Working Party [17] has suggested that these conditions are met with types II to VI of the subsoils set out in the table attached to Regulation D7 of the Building Regulations [11].

Granular bedding up to mid-diameter of the pipe (class 'B' bedding) is now used in the U.K. for most sewerage work when flexibly jointed pipes are used. A modified and stronger form of class 'B' bedding incorporates granular material to a level of 100 mm above the soffits of the pipes. The granular bedding material must first be laid to an approximate level and gradient; socket holes are then scooped out and each pipe is bedded into the material using the sight rails to obtain correct line and gradient as before. Heavy pipes should be suspended during laying and jointing to avoid disturbance of the granular bed, and to prevent the whole weight of the pipe from bearing on the rubber ring during jointing. Timbering or sheeting should, wherever possible, be withdrawn as work proceeds so that no voids are left in the bedding. There is a risk of the bedding acting as a permanent drainage channel for subsoil water and it may be necessary for waterstops to be constructed in some ground conditions to prevent this.

The Working Party [17] has recommended the use of a flat layer of selected granular bedding (in effect a 180° bedding) for certain conditions as an alternative method of bedding rigid pipes up to and including 225 mm diameter. They then suggest two further bedding factors:

    (i)   1·5 when the specified single-size granular material is employed; or

    (ii) 1·1 when all-in aggregate or sand is employed.

Concrete cradles or arches can be used where a higher bedding

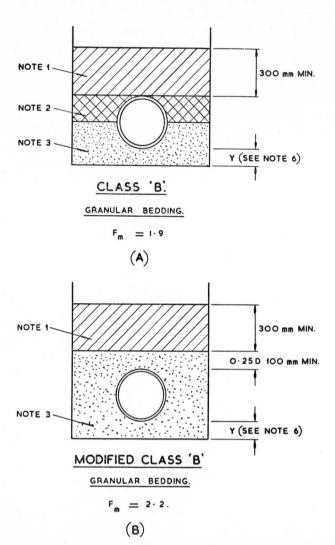

CLASS 'B'.

GRANULAR BEDDING.

$F_m = 1.9$

(A)

MODIFIED CLASS 'B'

GRANULAR BEDDING.

$F_m = 2.2$.

(B)

Fig. 10.1. Types of pipe beddings. Notes: (1) Filling material to be free from lumps, stones and roots; lightly compacted by hand. (2) Filling material to be free from lumps, stones and roots; carefully compacted around the pipes. (3) Granular bedding material. (4) Concrete with a minimum strength of $2 \cdot 1 \times 10^6$ kgf/m² (20·6 MN/m²) at 28 days. (5) 0·4% reinforcement for a bedding factor of 3·4, 1·0% reinforcement for a bedding factor of 4·8. (6) (a) In rock or mixed soils: $Y = 0 \cdot 25 \, B_c$ under barrels, with a minimum of 200 mm under both barrels and sockets. (b) In machine-dug uniform soils: $Y = 1/6 \, B_c$, with a minimum of 100 mm under both barrels and sockets. (c) In hand-shaped uniform soils: $Y = 100$ mm minimum under both barrels and sockets.

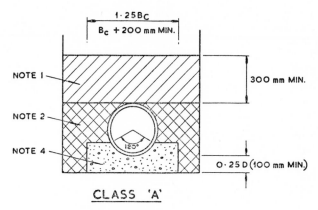

CLASS 'A'

PLAIN CONCRETE CRADLE.

$F_m = 2 \cdot 6$

(C)

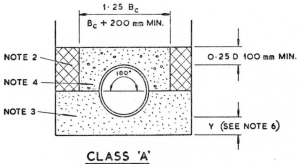

CLASS 'A'

PLAIN CONCRETE ARCH.

$F_m = 2 \cdot 6$

(D)

Fig. 10.1.—contd.

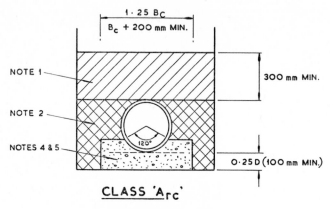

CLASS 'A$_{rc}$'

REINFORCED CONCRETE CRADLE.

$F_m = 3\cdot4$ OR $4\cdot8$.

(E)

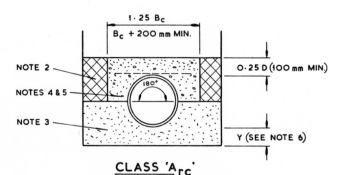

CLASS 'A$_{rc}$'

REINFORCED CONCRETE ARCH.

$F_m = 3\cdot4$ OR $4\cdot8$

(F)

Fig. 10.1.—contd.

factor is required; these can be either unreinforced or reinforced. A complete concrete surround can be used if a high strength bedding is necessary (see Table 10.2). Where a concrete protection is to be used (other than an arch) it may sometimes be convenient to put down a preliminary mat of weak concrete (about 50 mm thickness)

TABLE 10.2

UNREINFORCED CONCRETE SURROUND—BEDDING FACTOR 4·5

Nominal diameter of pipe (mm)	Dimensions of concrete surround	
	Under pipes and at sides (mm)	Over pipes (mm)
150	100	100
300	100	100
375	100	100
450	125	125
525	125	125
600	150	150
675	175	150
750	200	150
825	200	150
900	225	150
975	225	150
1 050	250	150

With acknowledgements to the National Clay Pipe Institute, U.S.A.

to form a clean working surface; this mat should be laid approximately to the required level and gradient. Concrete provides a positive, uniform and satisfactory bed for sewers, but it must be properly laid and of satisfactory quality; particular stress should be placed on the control of the quality of the concrete and the supervision of its placing.

Flexible pipelines must be given adequate bedding *and* side support, as a part of the vertical load will be transferred into a horizontal thrust, which must be resisted by the side fill. The usual practice is to bed *and surround* all flexible pipelines over 100 mm diameter with approved granular material to a depth of 150 mm over the pipes.

TESTING PIPELINES

Sewers with 'sealed' joints (including surface water sewers) should be tested after laying to ensure that they are sound, i.e. that no damaged pipes have been laid and that the joints are satisfactory. Sewers with ogee joints are not tested. Specifications should be so written that only requirements are included which are expected to be attained, and which it is intended to enforce; clauses which lay down absolute requirements which both the engineer and the contractor know to be unattainable in practice should be excluded.

C.P. 2005 recommends that tests for watertightness should be made on all lengths of sewer up to at least 760 mm diameter, together with all manholes and inspection chambers; many engineers insist on a further test *after* the trenches have been backfilled. Tests before backfilling should reveal faulty pipes and joints, while those carried out afterwards may disclose any faults in bedding or any subsequent damage during or after backfilling. If the full benefits of using mechanical joints are to be obtained, a quick and simple form of test is essential, and where there is any difficulty in obtaining water in sufficient quantity for a water test, many sewers are now tested with an air test. The air test is, however, not satisfactory for testing inverted siphons and other pipelines which will operate under pressure, nor can it be used after backfilling.

In the U.K., testing is normally carried out in accordance with conditions laid down in the Code [1]. With the water test, sewers will be subjected to an internal pressure test of 1·2 m head of water above the crown of the pipe at the high end (but not more than 6·0 m at the low end). After allowing a suitable time for absorption by pipes and joints, the loss of water is noted over a period of 30 min by adding water as necessary at 10-min intervals and noting the total quantity added. For smaller pipelines (up to 450 mm), the Code recommends that the loss should not be more than one litre *per hour* per linear metre per *metre* of nominal internal diameter. Pipelines over 450 mm diameter may be checked with a smoke test before being backfilled.

The air test is carried out under a test pressure of 100 mm of water, after air has been pumped in by a hand pump or similar method; the drop in pressure should not exceed 25 mm water head during a period of 5 min without further pumping.

The smoke test (referred to above) is carried out after both ends of the pipeline have been sealed with stoppers and smoke has been pumped into the line from a 'smoke machine'; any defective joint will be indicated by an escape of smoke. Smoke is pumped in at the lower end of the pipeline and the upper stopper should not be tightened until smoke appears at that end. This test can be used on existing drains as the pressure is insufficient to force smoke through a properly sealed trap.

A light test is used by many engineers to ensure that sewers have been laid to the correct line and gradient throughout any length between manholes. This can be carried out with a torch and one or two mirrors. A light is shone or reflected along the sewer from one end and, if the line is laid accurately to line and gradient, the light will be reflected at the mirror at the other end.

REINSTATEMENT OF TRENCH EXCAVATIONS

After thorough consolidation of the filling material up to 300 mm from the surface, trenches in carriageways should be filled with 225 mm of hardcore or lean mix road base material, followed by 75 mm of old road surface material or bitumen macadam, the whole being rolled with a 10-tonne roller until the surface is level with the surrounding carriageway surface. In footways, filling should be to within 150 mm of the surface and then completed with a 100 mm base of granular material or approved quarry waste plus 50 mm of bitumen macadam, the whole being rolled until level with the surrounding footway surface. Permanent reinstatement of carriageway and footway surfaces is normally carried out in the U.K. by the statutory highway authority.

BACKDROP AND SPECIAL MANHOLES

To avoid the cost of deep excavation for sewers with steep gradients, it may be more economical to lay a sewer at a gradient sufficient for the hydraulic requirements and then to connect this to any lower sewer by means of a backdrop manhole (see Fig. 10.2). The incoming sewer is led into a vertical pipe (normally constructed immediately outside the manhole), which in turn terminates at its lower end in a 90° bend at or just above the invert level of the lower sewer. Access is provided through the manhole wall to the upper level sewer for inspection or rodding.

A backdrop is often used when the difference in levels of the two sewers is more than about 600 mm; if the difference is less it can be taken up by using a ramp formed in the benching. For differences in level of up to about 1800 mm, many engineers prefer to use a 45°

Page 92, SURFACE WATER SEWERAGE.

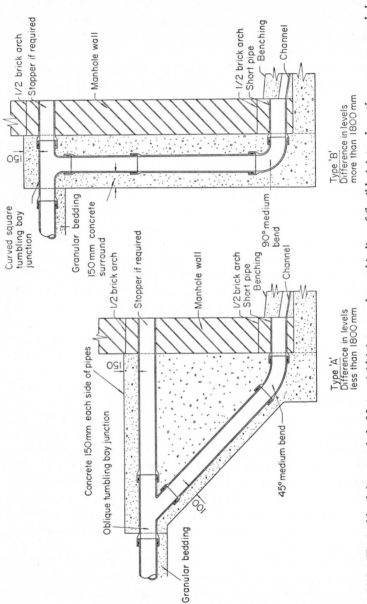

Fig. 10.2. *Typical backdrop manhole. Note: rigid joints may be used in lieu of flexible joints where these are surrounded with concrete.*

ramp, in the form of a 45° junction pipe turned with the branch feeding downwards, the main upper pipeline being extended forward into the manhole wall as before.

Attention has, however, recently been drawn to the cost of backdrop manholes, and where these are incorporated as a means of limiting the velocity in the sewer it is apparent that considerable saving is possible by *not* limiting the velocity. C.P. 2005 refers to experience in recent years which 'has shown that the scouring effect of sewage carrying road grit and other abrasive materials in suspension is less serious than was at one time thought', and the Code concludes that an upper limit of velocity to avoid scour is no longer of such great moment in design.

STRUCTURAL DESIGN

The subject of structural design of buried pipelines has been fully covered in various other reference works [35, 37, 38] and it is not intended to cover more than the basic principles in this chapter. The Marston Theory for structural design was set out in a National Building Studies Special Report [6] and a set of very useful tables was later published in 1970 [3]. This method is now an accepted part of sewerage design practice in the U.S.A., the U.K. and many other countries.

The usual assessment of loading conditions is based on the following:

(a) *For sewers laid under main traffic routes and under roads which are liable to be used for temporary diversion of heavy traffic,* provision should be made for a maximum of eight wheel loads, each of 9060 kg static weight, arranged as in B.S. 153 Type HB road loading, acting simultaneously with an impact factor of 1·3.

(b) *For sewers laid under other roads, except access roads used only for very light traffic,* provision should be made for a maximum of two wheel loads, each of 7250 kg static weight, spaced 900 mm apart, acting simultaneously with an impact factor of 1·5.

(c) *For sewers laid in fields, gardens and lightly trafficked access roads,* provision should be made for a maximum of two wheel loads, each of 3200 kg static weight, spaced 900 mm apart, acting simultaneously with an impact factor of 2·0.

(d) The maximum distributed surcharge load (e.g. stacks of earth, bricks, paving slabs, etc.) ever likely to be imposed on the ground over the pipeline should be assessed. If this is unlikely to be greater than 23 kN/m² (2400 kg/m²) over an area not larger than about 4·5 m square in plan, no separate provision need be made for it.

(e) The total effective design load will normally be the sum of the fill load plus the vehicle load.

(f) It will be possible in some cases, by limiting the width of the trench during construction, to design on narrow trench conditions.

(g) The bedding factors recommended in the National Building Studies Special Report [6] should be adopted.

To solve any specific problem, the following basic information is then required:

1. pipe diameter (nominal), in mm;
2. trench width for design purposes, in mm;
3. maximum trench depth, in mm;
4. minimum trench depth, in mm;
5. density of fill material, in kg/m³;
6. traffic loading and impact factor—see above.

As both trench width and pipe diameter have now been expressed in millimetres, the basic Marston formulae become:

Narrow-trench

$$W_c = C_d B_d^2 \times 10^{-6} \text{ kgf/m} \qquad \textbf{Formula 10.1}$$

Wide-trench

$$W_c' = C_c B_c^2 \times 10^{-6} \text{ kgf/m} \qquad \textbf{Formula 10.2}$$

Curves showing the values of coefficient C_d for various types of fill material are given in Fig. 10.3 [6]. The coefficient C_c is dependent on the ratio H/B_c and on the product of r_{sd} and p. Values of this coefficient for various values of $r_{sd}p$ are given in Fig. 10.4 [6]. Where no specific information on soil conditions is available, the values of γ and $K\mu'$ given in Table 10.3 are generally satisfactory.

Under wide trench conditions, for rigid pipe beddings (e.g. concrete) which do not extend over the full width of the trench, the settlement ratio (r_{sd}) and the projection ratio (p) may both be assumed as 1·0, giving a composite value of 1·0 for the expression

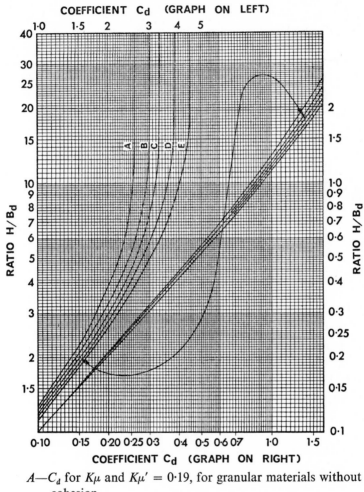

A—C_d for $K\mu$ and $K\mu' = 0\cdot19$, for granular materials without cohesion

B—C_d for $K\mu$ and $K\mu' = 0\cdot165$ max, for sand and gravel

C—C_d for $K\mu$ and $K\mu' = 0\cdot150$ max, for saturated top soil

D—C_d for $K\mu$ and $K\mu' = 0\cdot130$ ordinary max. for clay

E—C_d for $K\mu$ and $K\mu' = 0\cdot110$ max, for saturated clay

Fig. 10.3. Narrow trench fill load coefficents C_d (Crown copyright).

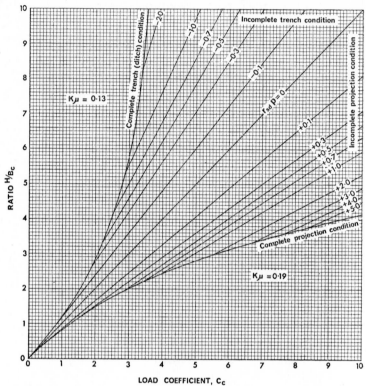

Fig. 10.4.   Wide trench fill load coefficients C_c *(Crown copyright).*

TABLE 10.3

SOIL DENSITIES AND COEFFICIENTS

Soil	Density (γ) (kg/m³)	Coefficient $K\mu'$
Sand and gravel	1 840	0·165
Saturated topsoil	1 920	0·150
Ordinary clay	2 000	0·130
Saturated clay	2 080	0·110

$r_{sd}p$. When granular bedding is to be used over the full width of a 'wide' trench, and up to the mid-diameter of a rigid pipe, however, the value of p may be taken as 0·5, so that $r_{sd}p$ then also becomes 0·5. When small diameter pipes (up to 300 mm diameter) are laid on the natural trimmed trench bottom, the value of $r_{sd}p$ should be taken as 0·7. For all wide trench conditions, K is generally taken as 0·190, as any lesser values would have very little effect on the value of coefficient C_c.

For any specific pipeline, it is now possible to calculate comparable values of W_c and W_c' using either Formula 10·1 or Formula 10·2,

TABLE 10.4
CONCENTRATED SURCHARGE LOADS

Bedding type	Traffic loading	Impact factor (F_i)	Graph
Class 'B'	Trunk roads	1·3	Fig. 10.5
or	Other roads	1·5	Fig. 10.6
concrete arch	Fields, etc.	2·0	Fig. 10.7
	Trunk roads	1·3	Fig. 10.8
Concrete cradle	Other roads	1·5	Fig. 10.9
	Fields, etc.	2·0	Fig. 10.10

along with suitable values of either C_d or C_c taken from Figs. 10.3 and 10.4. The *lower* of the two values (W_c or W_c') will be used in later calculations of the total loading on the pipeline.

The curves in Figs. 10.5 to 10.10 give values of concentrated surcharge loads (W_{csu}) as recommended by the Working Party. These curves have been developed from Charts C13 to C16 of the National Building Studies Special Report No. 37, and give metric values for the six possible conditions set out in Table 10.4. The curves in these six figures include allowances for the impact factors recommended by the Working Party.

The first set of calculations should be carried out, using Figs. 10.5, 10.6 and 10.7. If it is then found that a Class 'B' bedding is not adequate, the calculations must be repeated, using Figs. 10.8 to 10.10, if it is proposed to use a concrete cradle, and not an arch.

For larger diameter pipelines (over 300 mm diameter) it is usually

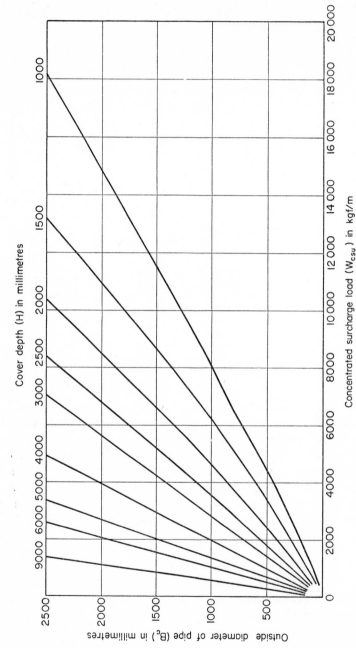

Fig. 10.5.   Concentrated surcharge loading—main traffic routes. Impact factor 1·3, Class 'B' bedding or concrete arch.

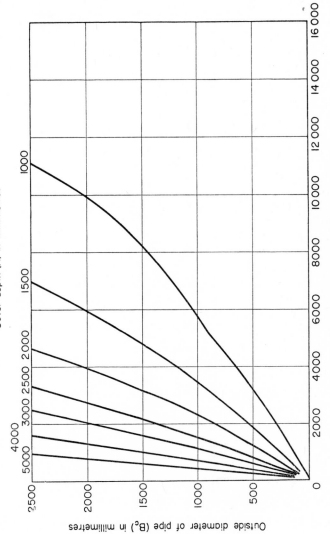

Fig. 10.6. Concentrated surcharge loading—other roads. Impact factor 1·5, Class 'B' bedding or concrete arch.

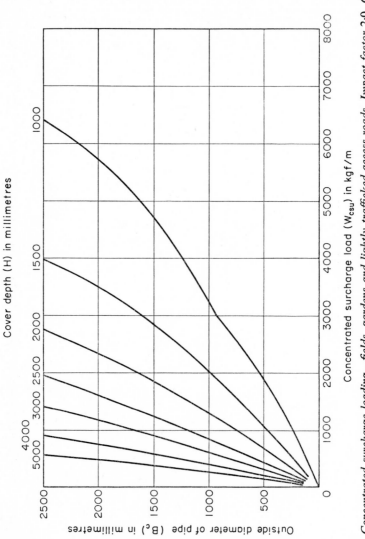

Fig. 10.7. Concentrated surcharge loading—fields, gardens and lightly trafficked access roads. Impact factor 2·0, Class 'B' bedding or concrete arch.

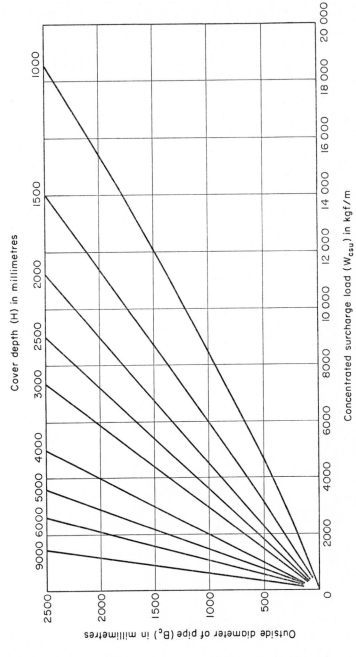

Fig. 10.8. Concentrated surcharge loading—main traffic routes. Impact factor 1·3, concrete cradle.

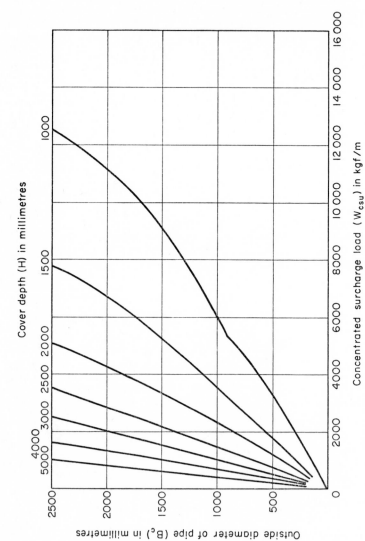

Fig. 10.9.　Concentrated surcharge loading—other roads. Impact factor 1·5, concrete cradle.

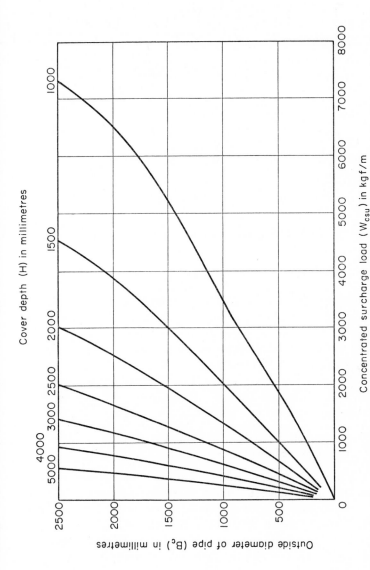

Fig. 10.10. Concentrated surcharge loading—fields, gardens and lightly trafficked access roads. Impact factor 2·0, concrete cradle.

advisable to add an additional allowance for the weight of the water in the pipes when calculating the total effective external load. Water weighs 1000 kg/m^3, and when the internal diameter is measured in millimetres the equivalent external load on rigid pipes can be taken for design purposes as

$$W_w = 0.75 \frac{\pi}{4} d^2 \times 10^{-3} \text{ kgf/m}$$

This can be simplified to

$$W_w = 589 d^2 \times 10^{-6} \text{ kgf/m} \qquad \textbf{Formula 10.3}$$

The total effective distributed load for design purposes (W_e) will then normally be

$\quad\quad W_c$ or W_c' (whichever is the lower)

plus

$\quad\quad W_{csu}$ $\quad\quad$ (the traffic loading)

plus

$\quad\quad W_w$ $\quad\quad$ (if applicable).

The strength of the pipeline itself (W_f) is calculated from the value of W_T (the safe crushing test strength of the pipes), multiplied by a suitable safety margin F_s when the pipes are unreinforced (recommended as 0·80 by the Working Party), and also multiplied by the bedding factor F_m (see Table 10.1).

The total effective distributed load on the pipeline (W_e) must, of course, be less than the strength of the pipeline (W_f). If the first calculation of W_f is too low, either:

(i) stronger pipes must be used (increase W_T); or
(ii) a stronger bedding must be specified (increase F_m).

If the proposed type of bedding is changed from granular (type 'B') to one of the concrete 'cradle' protections, the charts in Figs. 10.8 to 10.10 must be used to calculate a new value for W_{csu}; this may then entail a slight adjustment in the value of W_e.

When it is proposed to use an arch for Class 'A' or 'A_{rc}' bedding, two additional factors must be taken into account:

(i) the value of H must be taken as the height of cover over the concrete arch, and *not* over the pipe itself; and

(ii)   the value of B_c in wide trench calculations must be taken as
        the actual width of the top of the concrete, and *not* the out-
        side diameter of the pipes.

It will be apparent that item (i) above will normally have only a very
minor effect on the calculations, but that by increasing the value of
B_c the use of concrete arch protection under wide trench conditions
would increase the value of W_e. This will only be detrimental when
this would then make the value of W_e greater than W_f.

References

GOVERNMENT PUBLICATIONS
1. *Sewerage*, British Standard Code of Practice 2005.
2. *Drainage for Housing* (1966). Building Research Station Digest No. 6.
3. *Simplified Tables of External Loads on Buried Pipelines* (1970). Building Research Station, H.M.S.O.
4. *Charts for the Hydraulic Design of Channels and Pipes* (1969). Hydraulic Research Paper No. 2, H.M.S.O.
5. *Tables for the Hydraulic Design of Storm-drains, Sewers and Pipelines* (1969). Hydraulic Research Paper No. 4, H.M.S.O.
6. *Loading Charts for the Design of Buried Rigid Pipes* (1966). National Building Studies, Special Report 37, H.M.S.O.
7. *Taken for Granted* (1970). Report of the Working Party on Sewage Disposal, H.M.S.O.
8. *Rules for Rainfall Observers* (1965). H.M.S.O.
9. *Interim Report* (1963). Technical Committee on Storm Overflows and the Disposal of Storm Sewage, H.M.S.O.
10. *Final Report* (1970). Technical Committee on Storm Overflows and the Disposal of Storm Sewage, H.M.S.O.
11. *The Building Regulations* (1972). H.M.S.O.
12. *The Design of Urban Sewer Systems* (1962). Transport and Road Research Laboratory: Technical Paper 55.
13. *A Guide for Engineers to the Design of Storm Sewer Systems* (1963). Transport and Road Research Laboratory: Note No. 35.
14. *The Depth of Rain Water on Road Surfaces* (1968). Transport and Road Research Laboratory, Report LR 236.
15. *The Hydraulic Efficiency and Spacing of B.S. Road Gullies* (1969). Transport and Road Research Laboratory, Report LR 277.

16. *A Program for Calculating Thiessen Average Rainfall* (1972). Transport and Road Research Laboratory, Report LR 470.
17. *First Report* (1975). Working Party on Sewers and Water Mains, H.M.S.O.

TECHNICAL PAPERS

18. Ackers, P. *et al.* (1964). Effects of use on the hydraulic resistance of drainage conduits, *Proc. Inst. civ. Engrs.*, **28**, 339.
19. Bilham, E. G. (1935). Classification of heavy falls (of rain) in short periods, *British Rainfall*.
20. Braine, C. D. C. (1955). The effect of storage on sewerage design, *Proc. Inst. civ. Engrs.*, Pt III, **4** (2), 446.
21. Copas, B. A. (1957). Storm water sewer calculations, *J. Inst. P.H. Eng.*, **56**, 3, 137.
22. Davis, L. D. (1963). The hydraulic design of balancing tanks and river storage pounds, *J. Inst. Mun. Engrs.*, **90**, 1.
23. Escritt, L. B. (1946). Designing a surface-water sewerage system, *The Surveyor*, March 8.
24. Fortier, S. and Scobey, F. C. (1926). Permissible canal velocities, *Trans. Amer. Soc. Civ. Engrs.*, **89**, 940.
25. Hughes, T. P. (1969). Sewerage—a field for research, *J. Inst. Mun. Engrs.*
26. Lloyd-Davies, D. E. (1905–6). The elimination of storm-water from sewerage systems, *Proc. Inst. civ. Engrs.*, **164**(2), 41.
27. Mollinson, A. R. (1958). Road surface water drainage, *J. Inst. Highway Engrs.*, **V**, iv.
28. Ormsby, M. T. M. (1933). Rainfall and run-off calculations, *J. Inst. Mun. Engrs.*, **59**(16), 889.
29. Reid, J. (1927). The estimation of storm-water discharge, *J. Inst. Mun. Engrs.*, **53**, 997.
30. Riesbol, H. S. (1954). Snow hydrology for multi-purpose reservoirs, *Trans. Amer. Soc. Civ. Engrs.*, **119**, 595.
31. Rodda, J. C. (1967). The systematic error in rainfall measurement, *J. Inst. Wat. Engrs.*, **21**(2), 173.
32. Sharpe, D. E. and Shackleton, D. S. (1959). Balancing reservoirs—their use in surface water drainage schemes, *J. Inst. Mun. Engrs.*, **86**, 3, 80.
33. Wiesner, C. J. (1964). Hydrometeorology and river flood estimation, *Proc. Inst. civ. Engrs.*, **27**, 153.

TEXTBOOKS, ETC.

34. Barlow, T. (1926). *Hydraulics—Gauging of Sewage Flows, etc.,* Crosby Lockwood, St. Albans.
35. Bartlett, R. E. (1970). *Public Health Engineering: Design in Metric—Sewerage,* Applied Science Publishers, London.
36. Bartlett, R. E. (1971). *Public Health Engineering: Design in Metric—Wastewater Treatment,* Applied Science Publishers, London.
37. Clarke, N. W. B. (1968). *Buried Pipelines—A Manual of Structural Design and Installation,* Applied Science Publishers, London.
38. *Construction of Flexibly Jointed Concrete Pipelines* (1974). Concrete Pipe Assn., Bulletin No. 1.
39. Crimp and Bruges (1969). *Tables and Diagrams for Use in Designing Sewers and Water Mains,* Mun. Publications Co. Ltd., London.
40. Escritt, L. B. (1967). *Sewerage and Sewage Disposal, Calculations and Design,* Applied Science Publishers, London.

Definitions and Abbreviations

DEFINITIONS

Back-Drop Manhole:
: A manhole built at a junction of two sewers, where one sewer joins the other at a higher level and the sewage passes through a vertical or inclined shaft to the lower level.

Catchment Area:
: The area of a watershed discharging to a sewer, river or lake.

Combined Sewer:
: A sewer designed to carry both foul sewage and surface water.

Drain:
: Term generally relating to a channel or pipeline draining one building, or buildings, within the same curtilage.

Drainage Area:
: The area actually draining to a given point, which may or may not coincide with the 'catchment area'.

Gradient:
: The inclination of the invert of a pipeline expressed as a fall in a given length.

Hydraulic Gradient:
: The surface slope of a liquid in a pipeline. This is generally taken as parallel to the invert in a gravity sewer.

Impermeable Area:
: Often taken as the total of roofed and paved areas directly connected to a sewerage system.

Invert:
: The lowest point of the internal cross section of a sewer or channel.

Inverted Siphon:
: A portion of a pipe or conduit in which sewage flows under pressure, due to the

	sewer dropping below the hydraulic gradient and then rising again.
Manhole:	A chamber constructed on a sewer so as to provide access thereto for inspection, testing or the clearance of obstruction.
Rainfall:	Precipitation in any form, such as rain, snow, hail etc. The rate of rainfall is measured in millimetres per hour.
Run-off:	That part of rainfall which flows off the surface to reach a sewer or river.
Separate Sewer:	A sewer designed to carry foul sewage only or surface water only.
Sewage:	Water-borne human, domestic and farm waste. It may include trade effluent, sub-soil or surface water.
Sewer:	This term generally relates to any drain or sewer which drains more than one building, i.e. not legally a 'drain' as defined above.
Sewerage:	A system of sewers and ancillary works to convey sewage from its point of origin to a treatment works or other place of disposal.
Storm Sewage:	Foul sewage diluted with surface water.
Storm Sewage Overflow:	A weir, siphon or other device on a combined or partially separate sewerage system, introduced for the purpose of relieving the system of flows in excess of a selected rate, so that the size of the sewers downstream of the overflow can be kept within economical limits, the excess flow being discharged to a convenient watercourse.
Storm Tank:	A tank (generally located at a sewage treatment works) provided for storage and partial treatment of excess storm sewage before discharge to a watercourse.
Surface Water:	Natural water from the ground surface, paved areas and roofs.

Time of Concentration: The longest time taken for the rain falling on the drainage area to reach the point under consideration.

Trunk Sewer: A main sewer which takes the flow from a number of branch sewers, and serves as the main carrier of sewage for a large area.

ABBREVIATIONS

Cubic metre(s)	m^3
Cubic metres per second	m^3/s or cumec
Day(s)	d
Gravitational acceleration (9·806 m/s²)	g
Hectare (10^4 m²)	ha
Hour(s)	h
Kilogramme(s)	kg
Kilometre(s)	km
Metre(s)	m
Metres per second	m/s
Millimetre(s)	mm
Minute(s)	min
Newton(s)	N
Second(s)	s
Square metre(s)	m^2
Year (annum)	a

APPENDIX 'B'

Conversion Factors

Unit	Imperial		Metric
Length	0·621 4 miles	1	km 1·609
	3·281 ft	1	m 0·304 8
	1·094 yards	1	m 0·914 4
	0·049 7 chain	1	m 20·116 8
	0·547 fathom	1	m 1·829
	0·039 37 in	1	25·40 mm
	0·039 37 'thou'	1	μm 25·40
Area	$1·550 \times 10^{-3}$ in^2	1	mm^2 645·2
	10·764 ft^2	1	m^2 0·092 90
	1·196 yd^2	1	m^2 0·836 1
	0·386 1 sq mile	1	km^2 2·590
	2·471 acres	1	ha 0·404 7
Volume	35·315 ft^3	1	m^3 0·028 32
	1·308 yd^3	1	m^3 0·764 6
	$0·061 \times 10^{-3}$ in^3	1	mm^3 16 387·1
Capacity	0·220 Imp gal	1	litres 4·546
	0·264 2 U.S. gal	1	litres 3·785
	1·760 pints	1	litre 0·568
Velocity	3·281 ft/s	1	m/s 0·304 8
	196·8 ft/min	1	m/s 0·005 1
	0·621 4 m.p.h.	1	km/h 1·609
	0·539 6 knot (U.K.)	1	km/h 1·853 2
Mass	0·984 2 ton	1	tonnes 1·016
	0·019 7 cwt	1	kg 50·802
	2·205 lb	1	kg 0·453 6
	0·035 3 oz	1	g 28·349 5

113

Unit	Imperial		Metric
Mass/unit area	29·5 oz/yd^2	1	kg/m^2 33·90 × 10^{-3}
	0·001 4 lb/in^2	1	kg/m^2 703
	0·398 × 10^{-3} ton/acre	1	kg/ha 2 510·71
Rate of flow or	13·20 gal/min	1	litre/s 0·075 7
discharge	35·31 cusec	1	cumec 0·028 3
	2 118·6 cumin	1	cumec 0·47 × 10^{-3}
	19·01 m.g.d.	1	cumec 0·052 6
	3·675 gal/min	1	m^3/h 0·272
	219·97 gal/day	1	m^3/day 0·004 5
	14·3 ft^3/1 000 acres	1	litre/ha 0·070
	0·091 5 cusec/sq mile	1	litres/s km^2 10·933
Density	10·001 lb/gal	1	kg/litre 0·099 8
	0·062 43 lb/ft^3	1	kg/m^3 16·02
Force	0·224 8 lbf	1	N 4·448
	0·100 4 tonf	1	kN 9·964 0
	7·233 pdl	1	N 0·138 3
Force or weight per unit length	0·068 5 lbf/ft	1	N/m 14·593 9

Index

115